Assessment of the Operations of Osogbo
Transmission Line

Adewumi Adeloye

Assessment of the Operations of Osogbo-Akure 132 kV Transmission Line

A transmission line in the Nigerian Southwestern Grid is characterized and evaluated from multiple scientific dimensions

LAP LAMBERT Academic Publishing

Publisher:
LAP LAMBERT Academic Publishing
is a trademark of
International Book Market Service Ltd., member of OmniScriptum Publishing Group
17 Meldrum Street, Beau Bassin 71504, Mauritius

Printed at: see last page
ISBN: 978-613-4-99621-1

ASSESSMENT OF THE OPERATIONS OF OSOGBO-AKURE

132 kV TRANSMISSION LINE, NIGERIA

BY

ADELOYE, *ADEWUMI ADEOLUWA*

DEDICATION

This project is dedicated to the eternal, invisible God, from whom all wisdom comes.

ACKNOWLEDGEMENT

The accomplishment of this research will not have been possible without the assistance of certain important individuals. My profound gratitude goes to my supervisor, Dr A.O. Melodi, who exclusively supervised this project.

I also appreciate Engrs. Ugwute Francis and Adebiyi John who facilitated the process of acquiring logged data and transmission line parameters from the archives of Power Holding Company of Nigeria. They also assisted in the acquisition of the required quantities and unit prices of equipment for the preparation of my Bills of Engineering Measurement and Evaluation (BEME).

It is always a great pleasure to acknowledge with thanks the financial and moral support of my parents- Dr and Mrs V.S.A Adeloye-who entirely sponsored this degree programme.

ABSTRACT

The aim of this research was to assess the power transmission capability of the Osogbo-Akure 132 kV line in order to evaluate the existing performance indices of the transmission line, provide useful information for Akure district distribution system development and planning of its daily operations.

Circuit diagrams of Osogbo and Akure 132 kV switching substation along with the half-hourly active and reactive power components in MW and MVar respectively, for a period of 12 months, were collected from the National Control Centre, Osogbo (NCC). Descriptive statistics (maximum, mean, and standard deviations) of the half-hourly power data for the study period were evaluated in order to obtain half-hourly and daily maximum and minimum values in MW and MVar, which are relevant to power flow calculations in the line. The half-hourly phase angles (θ) and power factors for the study period were also determined concurrently. The daily load curve for each month and the study year were developed by plotting the half-hourly maximum statistics of data using Microsoft Excel spreadsheet. Power flow calculations were carried out using the method of successive or sequential approximations. A plot of the half-hourly statistical maximums of power (MW) obtained constituted the daily load curve. Furthermore, hourly load demand and the annual average costumer population data were obtained from the Akure end of the transmission line for a period of 60 months- 2005 to 2009. From the data obtained, the frequency and time durations of electric power outages were extracted and tabulated. The electric power outage data was used to compute the failure rate (λ) and the mean time between failures (MTBF) for each of the five years. A regression equation describing the trend depicted by the graph of failure rates and a formula for computing the amount of kilowatt of energy lost for each year were developed. Thereafter, the annual average customer population data and the values of the yearly failure rate were used to compute the customer oriented interruption indices.

A tee-off from the existing Osogbo-Benin 330 kV transmission line and a separate Osogbo-Akure 132 kV transmission line were considered as alternative sources of electric power supply for Akure district distribution network. The capacities of the transformer, circuit breaker and isolator required to be installed in the proposed substation were sized. A substation circuit diagram was designed for the two options and the Bills of Engineering Measurement and Evaluation (BEME) were produced. The bills were compared to identify the optimal option.

TABLE OF CONTENTS

LIST OF FIGURES

LIST OF TABLES

CHAPTER ONE

INTRODUCTION

1.1. PREAMBLE

The Electricity Corporation of Nigeria (ECN), a precursor of National Electric Power Authority (NEPA) -renamed Power Holding Company of Nigeria (PHCN) - was established as a statutory public corporation in 1951. The company took over power generation projects of the government which were carried out through the Public Works Department and from four native authorities. Between 1952/1953, the country generated 165 MW of power. Most of it was provided by ECN. During the following decade, the firm went through an expansion period increasing transmission lines available in Southern Nigeria; this was largely due to the rise in urbanization and demand for electricity. By 1964, the company had added additional power plants including one at Kano producing 6 megawatts of electricity and another at Ijora, Lagos producing 86.25 MW. It also opened new plants along the Oji river (25.5MW) and Afam (20MW) (Alex, 2008). An Eastern grid along Afam-Port Harcourt-Aba and Onitsha-Enugu-Nsukka with additional extensions at Nsukka, Calabar and Umuahia was also integrated to the National grid.

1.2. SIGNIFICANCE OF OSOGBO-AKURE 132 KV TRANSMISSION LINE

1964 was the same year that a western grid was created along Lagos-Ibadan-Ilorin with extensions at Abeokuta, Osogbo, Akure, Benin and Sapele. The Osogbo-Akure 132 kV annex of the western network spans 93 kilometres in route length. It passes through Kajola, Idominasi and Ilesha towns in Osun State. The transmission line continues through Igbara-Oke in Ondo State and eventually terminates in Akure, as can be observed in Fig 1.2. The conductor type is Aluminium Conductor Steel Reinforced (ACSR) and the entire length is paired. It is a single circuit radial line that is supported by 366 steel towers. This network provides bulk electric power to domestic commercial and industrial load points in Akure metropolis. The rate of production in the industries, general working condition and the living circumstances of the inhabitants of this reference region is enhanced through the installation of this extension of the national transmission network.

1

Figure 1.2 A section of the Nigerian Map

1.3.DEFICIENCIES IN THE NIGERIAN ELECTRIC POWER INDUSTRY

For many years, especially towards the end of the Nigerian second republic, the company was plagued by inefficiency in planning, management and maintenance and losses due to government debt and lack of proper pricing. Nigeria has 5900 MW of installed generating capacity. However, the country is only able to generate about 1900 MW because most facilities have been poorly maintained. The country has proven gas reserves and around 8000 MW of hydro development has been planned. Nigeria has plans to increase access to electricity throughout the country to 85% before the end of 2010. Achieving this feat would require 16 new power plants, approximately 15,000km of transmission lines, as well as distribution facilities. Nigeria's power sector has high energy losses (30 - 35 % from generation to billing), a low collection rate (75 - 80 %) and low access to electricity by the population (36 %) (Daniels, 2010). There is insufficient cash generation because of these inefficiencies and PHCN is consequently reliant on fuel subsidies and funding of capital projects by the government. At present only 20 % of rural households and 55 % of the country's total population have access to electricity (Alex, 2008). The Nigerian Energy Commission and the Solar Energy Society of Nigeria have been tasked with generating a solar-powered solution for the remote rural dwellers not served by the national power grid (Daniels, 2010).

PHCN's statutory obligations were rarely met before its extinction. Originally designed to be a self financing company remitting dividends to its owner and to provide constant electricity

2

to consumers and expand electricity provision to all local governments in Nigeria. Most of the financial and developmental goals were not met. It also had to battle with frequent collapses in its transmission lines leading to instability in its grid system and power outages. The inability of the firm to guarantee constant power supply has been a bottleneck in the manufacturing sector where many firms resorted to providing for their own power infrastructures. By the end of the 1980s, the company was only transmitting about half of its total installed capacity (Alex, 2008).

1.4. JUSTIFICATION OF RESEARCH

An assessment of the operation condition of the Osogbo-Akure 132 kV transmission network is vital for the following purposes:

a) identification of the current load limitations in the 132 kV power supply to Akure.

b) elimination or reduction of the identified load limitations to ensure an improved supply of electric power at nominal voltage level to Akure metropolis.

Most of the elements of the transmission section and substation of the grid e.g. transformers, protective devices and conductors, have been manufactured since 1964 (PHCN, 2005). Equipment as old as 44 years are prone to power quality problems as a result of deterioration and ageing. Within the last ten years, there were announcements of considerably heavy investment on the electric power industry without commensurable visible effects on power supply improvement to establish the veracity of the investments.

Load development, due to metropolitan developments and expansions since 1964, added to the concern on the adequacy of the study line. Furthermore, a physical examination of the Osogbo-Akure 132 kV transmission line reveals excessive vegetation growth along a considerable portion of its right of way (R-O-W). Very tall trees with extended branches were discovered almost throughout the entire portion of the R-O-W that passes through Kajola and Idominasi towns. Palm trees were also noticed within falling distance of the transmission line that transverses Osogbo and Ilesha towns.

3

1.5. AIM AND OBJECTIVES

The aim of this study is to assess the power transmission capability of the Osogbo-Akure 132 kV transmission line for improved power supply to Akure district distribution system.

The objectives of this study are to:

a. estimate the daily load curve of Osogbo-Akure 132 kV line;

b. determine the load current limitations of the Osogbo-Akure 132 kV transmission line;

c. assess the reliability of the Osogbo-Akure 132 kV transmission line; and

d. evaluate the possibility of feeding Akure from alternate source.

CHAPTER TWO

LITERATURE REVIEW

2.1   PREAMBLE

In this research, the sources selected for the review on the operational performance of transmission networks are textbooks on electric power principles, published reports of researches by experts in power systems and analysis in journals and the internet, and technical reports of foreign and local power system corporations. The intent of the review is to provide credible information on the following aspects: definitional information on power transmission networks and operations; performance evaluation methods in power transmission systems; and parameters of transmission networks and normal operating mode.

2.2   DEFINITIONAL INFORMATION

2.2.1   Transmission Network

Saccomanno (2003) described transmission networks, in the power supply system, as the element which transfers electric power from generating units to the distribution system. El-Hawary (2000) also defined a power transmission network as the main energy corridor through which power supply is being transferred to the customer through the distribution network.

A specimen of a transmission line is represented by section A-B in the Figure 2.1. Transmission lines, in the power supply system, transfers electric power from generating units to the distribution system which ultimately supplies the load. They also interconnect neighbouring utilities which allow the economic dispatch of power within regions during normal conditions, and the transfer of power between regions during emergencies.

5

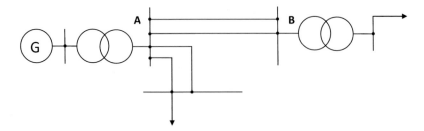

Figure 2.1 Electric Power System

The transmission section includes an extensive, relatively meshed network. A single generic line can carry hundreds or even thousands of megawatts, covering a more or less great distance, e.g., from 10 km to 1500 km and over. In Nigeria, transmission voltage lines operating at more than 33 kV are standardized at 132 kV and 330 kV. El-Hawary (2000) explained that high voltage transmission lines are terminated in substations, which are called high-voltage substations, receiving substations, or primary substations. The function of some substations is switching circuits in and out of service; they are referred to as switching stations. This was the opinion of Saccomanno (2003). At the primary substations, the voltage is stepped down to a value more suiTable for the next part of the trip toward the load. Very large industrial customers may be served from the transmission system. The portion of the transmission system that connects the high-voltage substations through step-down transformers to the distribution substations is called the sub transmission network. There is no clear distinction between transmission and sub transmission voltage levels. Sub transmission voltage level ranges from 33 kV to 132 kV. Some large industrial customers may be served from the sub transmission system.

Grigsby (2008) added that transmission lines can be either overhead, underground (cable) or submarine. There are high-voltage alternating current (HVAC) lines and high-voltage direct current lines (HVDC). Overhead transmission, sub transmission and primary distribution lines are strung between towers or poles. In urban settings underground cables are used primarily because of the impracticality of running overhead lines along city streets. However, underground cables are more reliable than overhead lines (because they have less exposure to climatic conditions such as hurricanes, ice storms, tornadoes, etc.). However, Dugan et. al. (2004) stated that underground cables are also much more expensive than overhead lines to

6

construct for unit of capacity and take much longer to repair because of the difficulty in finding the location of a cable failure and replacement.

From the definitions of authors sited above, it safe to conclude that the transmission network is a circuit that transfers generated electrical energy from power station to the load points through the distribution network.

2.2.2 Coincident and Non-Coincident Demand on Transmission Lines

Pabla (2005) defined demand as the average load over a specified time: often 15, 20, or 30 minutes. Demand can be used to characterize real power, reactive power, total power, or current. Meier (2006) concurs with Pabla's definition of demand and further explains that the peak demand over some period of time is the most common way utilities quantify a circuit's load. In substations, it is common to track the current demand.

In the utility context, Pabla presented a distinction between coincident and non-coincident demand. Coincident demand refers to the amount of combined power demand that could normally be expected from a given set of customers, say, a residential block on one distribution feeder. By contrast, the non-coincident demand is the total power that would be drawn by these customers if all their appliances were operating at the same time. Meier further explained that it is called non-coincident because all these demands do not usually coincide.

From the description and analysis presented by Pabla and Meier, one could conclude that demand (on transmission lines) is the average load over a specified time. Coincident demand reflects the statistical expectation regarding how much of these individual demands will actually overlap at any one time while the non-coincident demand is the total power that would be drawn by these customers if all their appliances were operating at the same time.

2.2.3 Electrical Load on Transmission Lines

Saccomanno (2003) stated that the electrical load on a transmission line connected to a feeder is the sum of all individual customer loads. Short (2004) has the same opinion with Saccomanno (2003) on the definitional depiction of electrical load on a transmission line. Short further explained that the electrical load of a customer is the sum of the load drawn by the customer's individual appliances. Customer loads have many common characteristics. Reeves and Heathcote (2003) described that load levels vary through the day, peaking in the afternoon or early evening. Allan (2009) in his opinion expressed that several definitions are used to quantify load characteristics at a given location on a circuit:

a) *Load factor* — The ratio of the average load over the peak load. Peak load is normally the maximum demand but may be the instantaneous peak. The load factor is between zero and one. A load factor close to 1.0 indicates that the load runs almost constantly. A low load factor indicates a more widely varying load. From the utility point of view, it is better to have high load-factor loads. Load factor is normally found from the total energy used (kilowatt-hours) as:

$$LF = \frac{kWh}{d_{kW} \times h}$$
2.1

where LF is load factor; kWh is energy use in kilowatt-hours; d_{kW} is peak demand in kilowatts; h is number of hours during the time period.

b) *Coincident factor* — The ratio of the peak demand of a whole system to the sum of the individual peak demands within that system. The peak demand of the whole system is referred to as the peak *diversified* demand or as the peak *coincident* demand. The individual peak demands are the *noncoincident* demands. The coincident factor is less than or equal to one. Normally, the coincident factor is much less than one because each of the individual loads do not hit their peak at the same time (they are not coincident).

$$CF = \frac{d_{kW}}{\Sigma_i \, dp_i}$$
2.2

where dp_i is i^{th} individual peak; CF is coincident factor

c) *Diversity factor* — The ratio of the sum of the individual peak demands in a system to the peak demand of the whole system. The diversity factor is greater than or equal to one and is the reciprocal of the coincident factor.

$$DF = \frac{\Sigma_i \, dp_i}{d_{kW}} = \frac{1}{CF}$$
2.3

$$DF \geq 1$$

d) *Responsibility factor* — The ratio of a load's demand at the time of the system peak to its peak demand. A load with a responsibility factor of one peaks at the same time as the overall system. The responsibility factor can be applied to individual customers, customer classes, or circuit sections.

$$RF = \frac{d_{L(t)}}{d_{kW(t)}}$$

2.4

where d_L is load demand; d_{kW} is peak load at time t

Reeves and Heathcote (2003) also stated that the loads of certain customer classes tend to vary in similar patterns. Allan (2009) shares a similar view with Reeves and Heathcote on the definitions used to quantify load characteristics at a given location on a circuit and the resemblance in the pattern of certain customer classes. Allan described that commercial loads are highest from 8 a.m. to 6 p.m. Residential loads peak in the evening. Weather significantly changes loading levels. On hot summer days, air conditioning increases the demand and reduces the diversity among loads. At the transformer level, load factors of 0.4 to 0.6 are typical.

The analysis and descriptions presented by the electric power system experts on the electrical loads on transmission lines could be accepted as reasonable.

2.2.4 Primary Components of an Overhead Transmission Line-HIGH VOLTAGE ALTERNATING CURRENT-HVAC (OVERHEAD).

El-Hawary (2000) declares that the primary components of an overhead transmission line are: conductors; ground or shield wires; insulators; support structures; and land or Right-of-Way (R-O-W).

a) Conductors: are the wires through which the electricity passes. Phase conductors in Extra-high Voltage (EHV) and Ultra-high Voltage (UHV) transmission systems employ aluminium conductors and aluminium or steel conductors for overhead ground wires. Many types of cables are available. These include:

1) Aluminium Conductors

There are five designs:

i. Homogeneous designs: These are denoted as All-Aluminium-Conductors (AAC) or All-Aluminium-Alloy Conductors (AAAC).

9

ii. Composite designs: These are essentially aluminium conductor-steel-reinforced conductors (ACSR) with steel core material.

iii. Expanded ASCR: These use solid aluminium strands with a steel core. Expansion is by open helices of aluminium wire, flexible concentric tubes, or combinations of aluminium wires and fibrous ropes.

iv. Aluminium-clad conductor (Alumoweld).

v. Aluminium-coated conductors.

Dugan et. al. (2004) shares a similar view with El-Hawary (2000) on the five designs of Aluminium conductors.

2) Steel Conductors

Galvanized steel conductors with various thicknesses of zinc coatings are used. Transmission wires are usually of the aluminium conductor steel reinforced (ACSR) type, made of stranded aluminium woven around a core of stranded steel which provides structural strength. When there are two or more of these wires per phase, they are called bundled conductors.

b) Ground or Shield Wires: are wires strung from the top of one transmission tower to the next, over the transmission line. Their function is to shield the transmission line from lightning strokes.

c) Insulators: are made of materials which do not permit the flow of electricity. They are used to attach the energized conductors to the supporting structures which are grounded. The higher the voltage at which the line operates, the longer the insulator strings. In recent years, polymer insulators have become popular in place of the older, porcelain variety. They have the advantage of not shattering if struck by a projectile. The most common form of support structure for transmission lines is a steel lattice tower, although wood H frames (so named because of their shape) are also used. In recent years, as concern about the visual impact of these structures has increased, tubular steel towers also have come into use.

d) Support Structure: The primary purpose of the support structure is to maintain the electricity carrying conductors at a safe distance from ground and from each other. Higher-voltage transmission lines require greater distances between phases and from the conductors to ground than lower-voltage lines and therefore they require bigger towers. The clearance from ground of the transmission line is usually determined at the midpoint between two successive towers, at the low point of the catenary formed by the line.

e) Right-of-way (R-O-W): The land that the tower line transverses is called the right-of-way (ROW). To maintain adequate clearances, as the transmission voltage increases, R-OW widths also increase. In areas where it is difficult to obtain R-O-Ws, utilities design their towers to carry multiple circuits. In many areas of the country it is not uncommon to see a structure supporting two transmission lines and one or more sub transmission or distribution lines. There are different philosophies on the selection of R-O-Ws. One philosophy is to try to site the corridor where there is little if any visual impact to most people. The other is that the R-O-W should be adjacent to existing infrastructure e.g. railroad, highways, natural gas pipelines. The intent of this is to minimize the overall number of corridors dedicated to infrastructure needs.

Cassaza and Delea (2003) align themselves with the opinion of El-Hawary (2000) on the primary components of an overhead transmission line-high voltage alternating current-HVAC. Cassaza and Delea further explains that reliability concerns argue for as much separation as possible between transmission R-O-Ws, to minimize exposure to incidents which might damage all lines on a R-O-W, so called common mode failures, such as ice storms, hurricanes, tornadoes, forest fires, airplane crashes, and the like. An ongoing issue with R-O-Ws is that they must be maintained to avoid excessive vegetation growth, which reduces the clearances between the line and ground.

2.2.5   Prevalent Causes of Electric Power Outages on Transmission Lines
Short-Circuit

Mehta and Mehta (2008) expressed that whenever a fault occurs on a network such that a large current flows in one or more phases, a short-circuit is said to have occurred. This is the cause of most of the electric power outage incidences on the transmission line in consideration. Lakervi and Holmes (1989) agree with the view of Mehta and Mehta on the description of short-circuit on a transmission line.

11

Causes of Short-Circuit

Lakervi and Holmes (1989) further explained that a short circuit in the power system is the result of some kind of abnormal conditions in the system. It may be caused as a result of internal and/or external effects. For instance:

i. Internal effects are caused by the breakdown of equipment or transmission lines, from deterioration of insulation in a generator, transformer etc. Such troubles may be due to ageing of insulation, inadequate design or improper installation.

ii. External effects causing short circuit include insulation failure due to lightning surges, over-loading of equipment causing excessive heating mechanical damage by public etc

Effects of Short-Circuit. When a short-circuit occurs, the current in the system increases to an abnormally high value while the system voltage decreases to a low value.

i. The heavy current due to short-circuit causes excessive heating which may result in fire or explosion. Sometimes, short-circuit takes the form of an arc and causes considerable damage to the system. For example, an arc on a transmission line not cleared quickly will severely burn the conductor causing it to break, resulting in a long time interruption of electric power supply.

ii. The low voltage created by the fault has a very harmful effect on the service rendered by the power system. If the voltage remains low, for even a few seconds, the consumers' inductive motors may be shut down and generators on the power system may become unstable.

Theraja and Theraja (2008) share the same opinion with Lakervi and Holmes (1989) on the causes and effects of short-circuit occurrence on transmission lines. Billinton & Allan (2008) recommends that due to the above detrimental effects of short-circuit, it is desirable and necessary to disconnect the faulty section and restore normal voltage and current as quickly as possible. Lakervi and Holmes agrees with the proposition of Billinton & Allan on the procedure of quickly restoring normal voltage and current in event of a short-circuit on a transmission line.

12

Other general causes of outage on the electric power transmission lines are presented by Theraja and Theraja as listed below:

1) Power utility's equipment failure
2) Costumer's equipment failure
3) Dig-in-for cables
4) Trees
5) Pollution
6) Storm
7) Flood
8) Lightning
9) Wear and tear
10) Accident
11) Power shortage
12) System inadequacy
13) Power theft
14) Lack of consumer care.

2.2.6   Operational Performance Evaluation

Nagrath and Kothari (2001) defined the operational performance evaluation of a network as an assessment of how well a network supports its function under normal and abnormal conditions. They further stated that the abnormal condition can either be introduced by failure of network elements or caused by events that generate abnormally high load shedding. Grigsby (2006) corroborated Nagrath and Kothari's opinion on the definition of the operational performance evaluation of an electric power grid and added that the emphasis of network performance evaluation is on assessment of the frequency of the abnormal conditions and the assessment of the impact on the consumers. The description of the evaluation of performance of electric transmission lines by Grainger and Stevenson (1994) was through the determination of the efficiency and voltage regulation of the line. Grainger and Stevenson further stated that the overall performance of a power system depends mainly on the optimal operation of the transmission lines.

Comparing the descriptions presented by the experts, it is safe to align ones opinion with that of Nagrath and Kothari. The reason for this choice is because his definition is exhaustive and has been substantiated by Grigsby.

2.3   FUNCTIONAL INFORMATION

2.3.1   Parameters of Transmission Networks and Normal Operating Mode

Grigsby (2006) explained that the working mode of 110-150 kV transmission lines, and by extension, the working mode of any electric power supply network is characterised by the parameters enumerated below:

Node voltage $V(V)$

Phase current $I(A)$

Frequency $f(Hz)$

Power flow $[\,S(MVA), P(MW), Q(MVar)\,]$

Power losses $[\,\Delta P(MW), \Delta Q(MVar)\,]$

Power factor $[\cos(\theta)]$

Weber (2005) identified the types of operating modes of an electric power system as enumerated below:

a) Normal steady state mode;

b) Abnormal steady state mode: permissible short duration overload;

c) Transient Mode: short circuit; voltage surges.

The values of each of these parameters must be constantly controlled so that they do not exceed safe limits (permissible maximum). In other words, $V_i \leq V_{max}$; $I_i \leq I_{permissible\ max}$, where V_i is the voltage of the ith bus and I_i is the current of the ith branch or power line. Based on the values of these quantities in a given time, the working mode of an electric power system can be classified into either the normal mode or fault mode. The normal mode is a working condition in which the regime parameters does not exceed safe limits while the fault mode is one in which the working mode exceed safe limits. An electric power system should possess high working reliability index. Furthermore, the economic parameters (total expenditure) on a network should be minimal without jeopardizing reliability and quality.

14

From the descriptions presented by the experts, one can accept the assertions of Grigsby and Weber on the definitions of the parameters of a transmission network and the normal system operating mode.

2.3.2 Parameters and Equivalent Circuits of Transmission Lines

Siemens (1998) stated that in order to calculate the load flow and other mode parameter in a network, it is necessary to represent each element of a one line diagram by an equivalent circuit diagram comprising the parameters of the elements. The presentation of an electric network in the form of a connected pool of equivalent circuits of separate elements of a network serves as the basis for the formation of the matrixes of their generalized parameter in a mathematical model. Cassaza and Delea (2003) and Short (2004) accept that this model sufficiently reflects the properties of these elements from the view point of the relationship between the mode parameters at the entrance and exit terminals of each element with the aid of unified set of parameters- impedances and admittances. The input data and the determination of the indicated parameters vary significantly for the different types of elements. For overhead lines, the parameters of the equivalent circuit are determined by the electrical conductivity of the conductor, the conductor size and the mutual arrangement on the pole support.

2.3.3 Parameters of Overhead (Air) Power Line

A power line (PL) is a part of an electric power system used for the transfer of electric energy from a source to a consumer. It is also used for intersystem connections. It is characterized by rated voltage V_r (V_N), section (or size, mm²) F, length $l(m)$, natural power transfer P_{nat}, and safe or maximum permissible load. The positive sequence parameters of the equivalent circuit of a line are generally determined by its length and specific (per-km) values of the active resistances and reactances, conductance and susceptance.

The values of the specific parameters of a line depends on factors such as the design number of circuits, number of conductors per phase, relative position of the phase and circuits and the material of the current carrying elements.

The overhead line parameters are as follows

2.3.3.1 Per-km impedance (z_0):

$$z_0 = r_0 + jx_0, \quad \text{[Ohm/km]} \qquad 2.5$$

15

where

$$r_0 : Per - km \; active \; resistance.$$

$$x_0 : Per - km \; reactive \; resistance.$$

2.3.3.2 Per-km admittance (y_0):
$$y_0 = g_0 + jb_0, \quad [sm/km] \qquad\qquad 2.6$$

where

$$g_0 : Per - km \; conductance.$$

$$b_0 : Per - km \; susceptance.$$

In the calculation of short and district networks, distribution of parameters is considered uniform.

2.3.3.3 Per-km active resistance (r_0):

Active resistance R_a is the opposition a conductor presents to the passage of alternating current (AC). Similarly, direct current resistance R_{dc} or R is the resistance of a conductor to direct current (DC). R_a is always greater than R_{dc} owing to skin effect. The tendency of an alternating current to concentrate near the surface of a conductor is known as skin effect Mehta and Mehta, 2008. The skin effect (SE) depends on the following factors:

a) Sectional area of the conductor *(F)*
b) Diameter *(d)* of the conductors   -increases with the diameter of wire.
c) Shape of wire   -less for stranded conductor than the solid conductor.

$$1.03 \leq \frac{R_a}{R_{dc}} \leq 1.05 \qquad\qquad 2.7$$

Line resistance: The resistance of a transmission line conductor is the most important cause of power loss in a transmission line. The resistance R_l of a line conductor having resistivity ρ (ohm.mm^2/km), length l (mm) and area of cross section F(mm^2) is given by;

$$R_l = \frac{\rho.l}{F} \qquad\qquad 2.8$$

$$= r_0.l$$

16

$$r_0 = \frac{1000}{\gamma F},$$

where

γ- Design conductivity [ohm.mm²/km]

The variation of resistance of metallic conductors with temperature is practically linear over the normal range of operation. Suppose R_1 and R_2 are the resistances of a conductor at $t_1°$C and $t_2°$C $(t_2 > t_1)$ respectively. If α_1 is the temperature coefficient at $t_1°$C, then,

$$R_2 = R_1[1 + \alpha_1(t_2 - t_1)] \qquad\qquad 2.9$$

where $\qquad \alpha_1 = \frac{\alpha_0}{1+ \alpha_0 t_1}$ (Dugan et.al. 2004)

$$\alpha_0 = temperature\ coefficient\ at\ 0°C$$

1) In a single phase or 2-wire d.c line, the total resistance (known as loop resistance) is equal to double the resistance of either conductor.

2) In case of a 3-phase transmission line, resistance per phase is the resistance of one conductor.

The value of R_a determines active power losses (ΔP_l) in a line. For a 3-phase line,

$$\Delta P_l = 3I^2R \qquad\qquad 2.10$$

$$= \frac{P^2 + Q^2}{V_N{}^2} . R_l \qquad\qquad 2.11$$

where $\qquad P = Active\ power\ component$

$$Q = Reactive\ power\ component$$

In steel conductors, there is skin effect (SE) and hysterisis. Therefore, $R_a > R_{dc}$

For steel, $r_0 = r_{0.dc} + r_{0.additional}$

where $\qquad r_{0.additional} = r_0^{SE} + r_0^{hysteresis}$

2.3.3.4 Per-km Inductive reactance of a power line (x_0):

$$x_0 = 0.144\ log\left(\frac{D_{GMD}}{r}\right) + 0.016,\ [Ohm/km] \quad (Reeves\ \&\ Heathcote, 2003)\ 2.12$$

where

$r = radius\ of\ conducor\ (taken\ from\ the\ technical\ directory)$

$D_{GMD} = geometric\ mean\ distance\ between\ phases\ (D_{GMD} = \sqrt[n]{D_1. D_2 \dots D_n})$

For a three phase single circuit line,

$$D_{GMD} = \sqrt[3]{D_{12}. D_{13}. D_{23}}\ between\ phases\ A, B\ and\ C.$$

It is possible to calculate the line reactive resistance, $X_l = x_0 l$ if given the length l of the line.

2.3.3.5 Reactive conductivity or line susceptance (B_l)

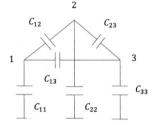

$$C_o = \frac{0.024}{log\frac{D_{GMD}}{r}}\ 10^{-6} \qquad [F/km] \qquad\qquad 2.13$$

Per-kilometre, $b_o = \omega C_o$

$$b_o = \frac{7.58}{log\frac{D_{GMD}}{r}}\ 10^{-6} \qquad [Sm/km] \qquad\qquad 2.14$$

Splitting of line leads to an increase in line capacitance C_o and line susceptance B_l accordingly. However, conductor splitting causes a decrease in corona losses.

2.3.4 Mathematical Modelling Methods of Electric Power Networks for Numerical Analysis

Nagrath and Kothari (2001) explained that the transmission lines are modelled using four parameters which affect its performance characteristic. These parameters are identified as series resistance, series inductance, shunt capacitive and shunt conductance. Saadat (2002)

shared the same opinion with Nagrath and Kothari on the parameters used for modelling of transmission lines.

The most commonly used circuit representations of transmission line are:

1. Nominal T circuit.
2. Nominal π circuit.

2.3.4.1 Nominal T circuit

Gupta (2008) presents a nominal T circuit of a transmission line in Figure 2.2 where the total capacitance of the conductor is concentrated at the centre of the line and the series impedance is split into two equal parts.

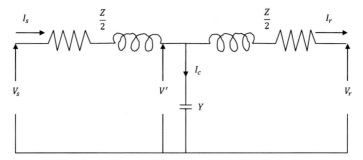

Fig 2.2 Nominal T circuit of a medium length transmission line

$$V' = V_r + I_r \frac{Z}{2} \qquad\qquad 2.15$$

$$I_c = V'Y = \left(V_r + I_r \frac{Z}{2}\right)Y \qquad\qquad 2.16$$

$$I_s = I_r + I_c = I_r + V_rY + I_r\frac{ZY}{2} = V_rY + I_r\left(1 + \frac{ZY}{2}\right) \qquad\qquad 2.17$$

$$V_s = V' + I_s\frac{Z}{2} = V_r + I_r\frac{Z}{2} + (V_rY + I_r + I_r\frac{ZY}{2})\frac{Z}{2}$$

$$V_s = V_r\left(1 + \frac{ZY}{2}\right) + I_rZ\left(1 + \frac{ZY}{4}\right) \qquad\qquad 2.18$$

Equations 2.13 and 2.14 can be written in the matrix form as:

19

$$\begin{bmatrix} V_s \\ I_s \end{bmatrix} = \begin{bmatrix} 1 + \dfrac{ZY}{2} & Z\left(1 + \dfrac{ZY}{4}\right) \\ Y & 1 + \dfrac{ZY}{2} \end{bmatrix} \begin{bmatrix} V_r \\ I_r \end{bmatrix}$$ 2.19

2.3.4.2 Nominal π circuit

Gupta (2008) also presents a nominal π circuit of a medium length transmission line in Figure 2.2 where half the total capacitance of each conductor is concentrated at the sending-end of the radial line, while the other half of it is at the receiving end the transmission line.

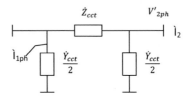

Fig 2.3 Nominal π circuit model of a transmission line

The circuit in Fig 2.3 can be characterized in two modes of the π -form circuit: no–load (NL) and short circuit (SC).

1) NL Mode: $V_{2\phi}, I_2 = 0$

$$V_{1ph} = V_{2ph} + I_L Z_{cct}$$ 2.20

$$I_L = I_2 + I_{ph2}$$ 2.21

$$I_L = V_{2ph} \frac{Y_{cct}}{2}$$

$$V_{1ph} = V_{2ph} + V_{2ph} Z_{cct} \frac{Y_{cct}}{2}$$

$$V_{1ph} = V_{2ph} \left(1 + Z_{cct} \frac{Y_{cct}}{2}\right)$$ 2.22

$$A = ch\sqrt{ZY}$$

$$= 1 + Z_{cct} \frac{Y_{cct}}{2}$$

$$I_1 = I_L + I_{Y1}$$

$$= V_{2ph} \frac{Y_{cct}}{2} + V_{2ph} \left(1 + Z_{cct} \frac{Y_{cct}}{2}\right) \frac{Y_{cct}}{2}$$

$$= V_{2ph}Y_{cct} + V_{2ph}Z_{cct}\frac{Y_{cct}^2}{4}$$

$$I_1 = V_{2ph}Y_{cct}\left(1 + Z_{cct}\frac{Y_{cct}}{2}\right) \qquad\qquad 2.23$$

$$C = \frac{1}{\sqrt{Z}}sh\sqrt{ZY}$$

$$= Y_{cct}(1 + Z_{cct}\frac{Y_{cct}}{4})$$

2) SC Mode: $V_{2\Phi} = 0, I_2$.

$$I_L = I_2 + I_{2ph} \qquad\qquad 2.24$$

$$I_L = I_2$$

$$V_{1ph} = V_{2ph} + I_L Z_{cct} \qquad\qquad 2.25$$

$$= I_2 Z_{cct}$$

$$B = \sqrt{\frac{Z}{Y}}sh\sqrt{\frac{Z}{Y}}$$

$$= Z_{cct}$$

$$I_1 = I_L + I_{Y1}$$

$$= I_2 + I_2 Z_{cct}\frac{Y_{cct}}{2}$$

$$I_1 = I_2(I + Z_{cct}\frac{Y_{cct}}{2}) \qquad\qquad 2.26$$

Gupta (2008) substantiates the opinion of Nagrath and Kothari (2001) on the effect of shunt capacitance becoming more pronounced with the increase in the length of the line, and shunt capacitance being considered as lumped in lines between 80 and 250 km in length.

Depending upon the manner in which the capacitance of the line is being considered, Gupta also described the characteristics of a medium length transmission line as stated: When the length of an overhead transmission line is about 50-150km and the line voltage is moderately high ($20kV < V_l < 100kV$), it is considered as a medium transmission line. Due to sufficient length and voltage of the line, the capacitance effects are taken into account. For purposes of calculations, the distributed capacitance of the line is divided and lumped in the form of condensers shunted across the line at one or more points.

The details presented on the characteristics of medium length transmission line were also corroborated by Grainger and Stevenson (1994). For this reason, the descriptions made by the specialists on the mathematical modelling methods of power networks for numerical analysis and the features of a medium length transmission line were accepted as valid.

2.4   LOAD FLOW

According to Grainger and Stevenson (1994), the load flow study in a power system constitutes a study of paramount importance. The study reveals the electrical performance and power flows (real and reactive) for specified conditions when the system is operating under steady state. The load flow study also provides information about the line and transformer loads (as well as losses) throughout the system and voltages at different points in the system for evaluation and regulation of the performance of the power system under conditions known as priori. Further alternative plans for the future expansion to meet new load demands can be analysed and complete information is made through this study.

Grigsby (2006) stated that the goal of a power flow study is to obtain complete voltage angle and magnitude information for each bus in a power system for specified load and generator real power and voltage condition. Once this information is known, real and reactive power flow on each branch as well as generator reactive power output can be analytically determined. Due to the nonlinear nature of this problem, numerical methods are employed to obtain a solution that is within an acceptable tolerance limit.

The solution to the power flow problem begins with identifying the known and unknown variables in the system. The known and unknown variables are dependent on the type of bus. A bus without any generators connected to it is called a Load Bus. With one exception, a bus with at least one generator connected to it is called a Generator Bus. The exception is one arbitrarily-selected bus that has a generator. This bus is referred to as the Slack Bus.

In the power flow problem, it is assumed that the real power P_D and reactive power Q_D at each Load Bus are known. For this reason, Load Buses are also known as PQ Buses. For Generator

22

Buses, it is assumed that the real power generated P_G and the voltage magnitude $|V|$ is known. For the Slack Bus, it is assumed that the voltage magnitude $|V|$ and voltage phase θ are known. Therefore, for each Load Bus, the voltage magnitude and angle are unknown and must be solved for; for each Generator Bus, the voltage angle must be solved for; there are no variables that must be solved for the Slack Bus.

2.5   POWER FLOW METHODS

2.5.1   Gauss-Seidel Method for Load Flow Studies

Gauss-Seidel (G-S) method is one of the most common methods used in load flow studies. The advantages of the method are:

a)   The simplicity of the technique;

b)   Small computer memory requirement;

c)   Less computational time per iteration.

The disadvantages of this method are:

a)   Slow rate of convergence and, therefore, large number of iterations.

b)   Increase of number of iterations directly with the increase in the number of buses.

c)   Effect on convergence due to choice of slack bus.

In view of these disadvantages, *G-S* method is used only for systems having small number of buses.

1) G-S Method when PV Buses are Absent

The discussion of *G-S* algorithm can be approached from a situation in which a system in which voltage controlled buses are absent. This implies that out of the n buses, one bus is slack bus and the remaining n-1 buses are *P-Q* buses. Initially, the magnitudes and angles at these n-1 buses are assumed. The voltages are then updated at every step of iteration.

From equation (2.27) $I_i = Y_{ii}V_i + \sum_{\substack{p=1 \\ p \neq 1}}^{n} Y_{ip}V_p$ 2.27

   or

$$V_i = \frac{1}{Y_{ii}}\left[I_i - \sum_{\substack{p=1 \\ p \neq 1}}^{n} Y_{ip}V_p\right]$$ 2.28

2.5.2 Newton–Raphson Method for Load Flow Studies

Newton-Raphson (NR) method is very suiTable for load flow studies on large systems. The advantages of this method are:

a) More accuracy and surety of convergence

b) Only about three iterations are required as compared to more than 25 or so as required by G-S method.

c) The number of iterations is almost independent of the system size

d) This method is insensitive to factors like slack bus selection, regulating transformers, e.t.c.

The disadvantages of this method are:

a) The solution technique is difficult

b) The calculation in each iteration procedure is more. Thus, computer time per iteration is large.

c) The computer memory requirement is large.

NR method can be applied to load flow problems in a number of ways. It is commonly used in solving rectangular and polar co-ordinates (Saadat, 2002; Grainger & Stevenson, 1994).

In this formulation, the quantities are expressed in rectangular form,

Let
$$V_p = |V_p| \angle \delta_p = e_p + jf_p$$

The active and reactive power components at each bus i_p are functions of e and f. Thus,

$$P_i = u_1 (e, f) \qquad\qquad\qquad 2.29$$

$$Q_i = u_2 (e, f) \qquad\qquad\qquad 2.30$$

For a system of n buses and bus 1 designated as slack bus, the difference equations which relate the changes in active and reactive power to changes in e and f take the form, for the ith bus,

$$\Delta P_i = \sum_{p=2}^{n} \frac{\partial P_i}{\partial e_p} \Delta e_p + \sum_{p=2}^{n} \frac{\partial P_i}{\partial f_p} \Delta f_p \qquad\qquad 2.31$$

24

$$\Delta Q_i = \sum_{p=2}^{n} \frac{\partial Q_i}{\partial e_p} \Delta e_p + \sum_{p=2}^{n} \frac{\partial Q_i}{\partial f_p} \Delta f_p \qquad 2.32$$

ΔP_i and ΔQ_i represent the difference between the specified and the calculated values of P_i and Q_i respectively. There are two equations similar to equations (2.31) and (2.32) for each of the $(n - 1)$ buses. In matrix form, these equations can be written as:

$$\begin{bmatrix} \Delta P \\ \Delta Q \end{bmatrix} = \begin{bmatrix} J_1 & J_2 \\ J_3 & J_4 \end{bmatrix} \begin{bmatrix} \Delta e \\ \Delta f \end{bmatrix} \qquad 2.33$$

Where J_1, J_2, J_3, J_4, are the elements of the Jacobian and are partial derivatives similar to those appearing in equations (2.31 and 2.32).

2.5.3 Fast Decoupled Load Flow

Load flow studies can be made faster and more efficient when the advantage of the physical properties of the system is taken into consideration. One such property is the network sparsity. Since each bus is connected to only a small number (usually two or three) of other buses, Y_{bus} of a large network is very sparse (i.e., it has a large number of zero elements). The sparsity feature of Y_{bus} minimises the computer memory requirement (because only non-zero terms need to be stored) and results in faster computations. Another property is the loose physical interaction between MW and MVar flows in a power system. Therefore, the MW-δ and MVar-V calculations can be decoupled. This decoupling results in a very simple, fast and reliable algorithm. The accuracy is comparable to that of NR method. (Saccomanno, 2003; El-Hawary, 2000).

2.5.4 Method of Sequential or Successive Approximations

The approach of successive approximation to load flow study is that it has provisions for direct computation of unknown values of PQ. It does not require any iterations; therefore,

a) The technique is simple,

b) Small computer memory is required,

c) The results obtained are very accurate.

The main disadvantage of the load flow technique becomes evident when numerous buses are being considered in a power system. The unknown variables for each bus have to be calculated individually and independently. As a result, calculation of values of unknown variables for a power system with n>1 buses may be slow.

25

1. Application to Transmission Network

In transmission networks, network mode is computed in two stages:

a. First approximation: power distribution and losses in the network are determined. The losses in the section of the transmission line can be computed from equations (2.34) and (2.35).

$$\Delta P_l = \frac{P_2^2 + Q_2^2}{V_2^2} r \approx \frac{P_1^2 + Q_1^2}{V_1^2} r \qquad \text{2.34}$$

$$\Delta Q_l = \frac{P_2^2 + Q_2^2}{V_2^2} x \approx \frac{P_1^2 + Q_1^2}{V_1^2} x \qquad \text{2.35}$$

where r and x are the load resistance and reactance respectively

P, Q and V are as previously defined.

In approximate calculation of power losses in the equations (2.34) and (2.35), the nominal voltage V_N is substituted for the actual voltages at the beginning and end of the line.

b. Second approximation: the determination of node voltages if the sending end voltage is given.

In conducting the second approximation, the successive determination of voltage losses in every section of the network is begun from the node where voltage is given. The voltage V_1 at the beginning of a line section, having active resistance r and reactance x during active power transfer, is determined, given powers and voltage $P_2, Q_2, and\ V_2$, at the end of the section by the equation (2.36):

$$V_1 = \sqrt{\left(V_2 - \frac{P_2 r + Q_2 x}{V_2}\right)^2 + \left(\frac{P_2 x - Q_2 r}{V_2}\right)^2} \qquad \text{2.36}$$

or

$$V_1 = \sqrt{(V_2 - \Delta V)^2 + (\delta V)^2} \qquad \text{2.37}$$

where ΔV and δV are the direct and quadrature components

of voltage drop in the given section

Analogically, V_2 at the end of the section can be determined, given data at the beginning of the section by the equation (2.38).

26

$$V_2 = \sqrt{\left(V_1 - \frac{P_1 r + Q_1 x}{V_1}\right)^2 + \left(\frac{P_1 x - Q_1 r}{V_1}\right)^2}$$ 2.38

2.6   CONCLUSION OF REVIEW

This literature review has attempted to provide reliable information on the definitional explanation on: power transmission networks and operations; performance evaluation methods in power transmission systems; parameters of transmission networks and normal operating mode.

Transmission network is that part of a power system that transfers generated electrical energy from power station to the load points through the distribution network. The operational performance evaluation of a network is an assessment of how well a network supports its function under normal and abnormal conditions. An abnormal condition can be introduced by failure of network elements and surges produced from lightening strokes.

The working mode of 110-150 kV transmission lines, and by extension, the working mode of any electric power supply network is characterised by the parameters enumerated below:

Node voltage $V(V)$

Phase current $I(A)$

Frequency $f(Hz)$

Power flow $[\, S(MVA), P(MW), Q(MVar)\,]$

Power losses $[\, \Delta P(MW), \Delta Q(MVar)\,]$

Power factor $[\cos(\theta)]$

The types of operating modes of an electric power system are as enumerated below:
  a) Normal steady state mode;
  b) Abnormal steady state mode: permissible short duration overload;
  c) Transient Mode: short circuit; voltage surges.

The load flow study in a power system constitutes a study of paramount importance. The study reveals the electrical performance and power flows (real and reactive) for specified conditions when the system is operating under steady state. The load flow study also provides information about the line and transformer loads (as well as losses) throughout the system and voltages at different points in the system for evaluation and regulation of the performance of the power system under conditions known as priori. Further alternative plans for the future expansion to meet new load demands can be analysed and complete information is made through this study.

In this research, the method of sequential or successive approximations was adopted for the load flow study of the Osogbo-Akure 132 kV transmission line. This is because it has provision for direct computation of unknown values of PQ. The transmission line that this research focuses on is just one. Therefore, no iterative process is required in the computation. The sequential technique is simple, a very small computer memory is required and the results obtained are accurate.

CHAPTER THREE

METHODOLOGY

3.1   ESTIMATION OF THE DAILY LOAD CURVE OF OSOGBO-AKURE 132 KV
LINE.

3.1.1   Acquisition of Circuit Diagrams and Half-hourly Power Components

The circuit diagrams of the Osogbo and Akure 132 kV switching substations and the half-
hourly active and reactive power components for a period of twelve months were obtained
from the National Control Centre, Osogbo. The substation circuit diagrams obtained are
presented in figures 3.1 and 3.2 while the power components supplied for every half hour of
each day in the month was tabulated as shown in Table 3.1.

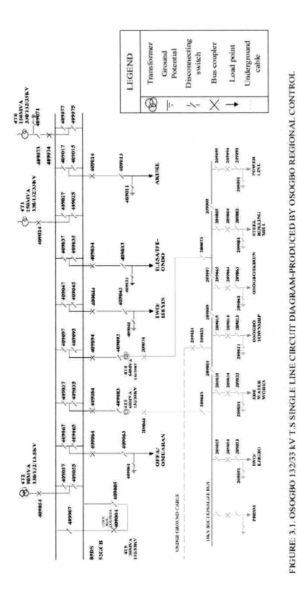

FIGURE: 3.1. OSOGBO 132/33 kV T.S SINGLE LINE CIRCUIT DIAGRAM-PRODUCED BY OSOGBO REGIONAL CONTROL

Source: Archives of National Control Centre, Osogbo, Osun State, Nigeria.

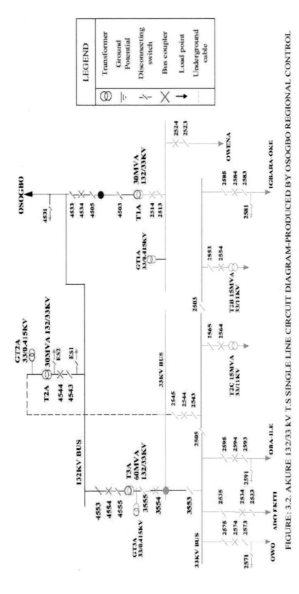

FIGURE: 3.2. AKURE 132/33 kV T.S SINGLE LINE CIRCUIT DIAGRAM-PRODUCED BY OSOGBO REGIONAL CONTROL

Source: Archives of National Control Centre, Osogbo, Osun State, Nigeria.

TABLE 3.1   A half-hourly daily load record for 9 May 2009.

Time	Voltage	Active Power	Reactive Power
Minutes	(KV)	(MW)	(MVar)
00:00	135	34	15
00:30	135	30	14
01:00	138	31	12
01:30	136	31	13
02:00	136	32	14
02:30	137	31	14
03:00	137	32	13
03:30	137	33	14
04:00	136	32	13
04:30	131	31	14
05:00	135	13	15
05:30	136	35	16
06:00	135	38	17
06:30	136	37	16
07:00	137	35	15
07:30	137	36	15
08:00	136	30	13
08:30	135	32	16
09:00	136	37	20
09:30	136	37	20
10:00	136	18	8
10:30	137	18	8
11:00	135	18	9
11:30	135	18	9
12:00	135	18	9

12:30	135	18	8
13:00	134	16	9
13:30	135	16	6
14:00	135	16	6
14:30	135	16	7
15:00	134	32	16
15:30	134	32	16
16:00	131	42	25
16:30	132	39	21
17:00	133	26	10
17:30	133	26	10
18:00	133	26	10
18:30	135	30	11
19:00	137	32	11
19:30	136	29	9
20:00	136	23	8
20:30	134	24	8
21:00	136	26	8
21:30	134	28	9
22:00			
22:30	135	28	9
23:00	136	30	9
23:30	138	31	10
24:00	139	32	11

Source: Archive of Power Holding Company of Nigeria (PHCN) 132 kV Switching

Substation, Osogbo, Osun State, Nigeria.

3.1.2. Generation of Daily Load Curve

Consequently, a daily load curve was plotted from the tabulated data in order to examine the consistency in power transmission every 24 hours. A sample can be observed in figure 3.3 where the gaps represent incidences of electric power failure.

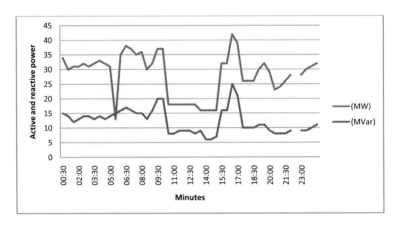

FIGURE 3.3 Graph of daily half hourly Readings of Load at 132 kV for 9 May 2009.

Microsoft Office Excel 2007® was used for the computation of the phase angle (θ) and power factor (pf) for each day in the 12 months duration. These values were computed from the relationship represented in equations (3.1) and (3.2) respectively.

$$(\theta_{i,j}) = tan^{-1} \frac{Max\ Q_{i,j}}{Max\ P_{i,j}}$$ 3.1

where $Q_{i,j}$ and $P_{i,j}$ represent active and reactive power in time i and day j

$$pf_{i,j} = cos\ (phase\ angle) = cos(\theta_{i,j})$$ 3.2

Afterwards, the mean and standard deviation (standard error) of active and reactive power components for every 30 minutes in each day of the 12 months were also calculated from using equations (3.3) and (3.8) and tabulated in Table 4.3.

$$P_{av\ i,j}^{(30)} = \frac{\sum_{\substack{j=1 \\ i=1}}^{N} P_{i,j}}{N}\ [MW]$$ 3.3

34

$$Q_{av\,i,j}^{(30)} = \frac{\sum_{\substack{j=1 \\ i=1}}^{N} Q_{i,j}}{N} \quad [MVar] \qquad\qquad 3.4$$

where $P_{av\,i,j}^{(30)}$ and $Q_{av\,i,j}^{(30)}$ are half-hourly mean values of active and reactive power components.

$$\sigma_{P(i,j)} = \sqrt{\left\{ \frac{\sum_{\substack{j=1 \\ i=1}}^{N} f(P_{i,j}\,(MW) - \bar{P}_{i,j})^2}{N} \right\}} \qquad\qquad 3.5$$

$$\sigma_{Q(i,j)} = \sqrt{\left\{ \frac{\sum_{\substack{j=1 \\ i=1}}^{N} f(Q_{i,j}\,(MVar) - \bar{Q}_{i,j})^2}{N} \right\}} \qquad\qquad 3.6$$

where σ_P and σ_Q are standard deviations of active and reactive power

from their mean values $\bar{P}, \bar{Q}$

$$Mean\ active\ power\ (\bar{P}_{i,2009}) = \frac{\sum_{\substack{j=1 \\ i=1}}^{N} P_{i,j}}{N} \qquad\qquad 3.7$$

$$Mean\ reactive\ power\ (\bar{Q}_{i,2009}) = \frac{\sum_{\substack{j=1 \\ i=1}}^{N} Q_{i,j}}{N} \qquad\qquad 3.8$$

where

f is : frequency of each value observed

N is : Number of observations of $P_i(MW)$ and $Q_i(MVar)$

j is : {1,2,3, … … … … … … … … … … … … ..,365} days in a year

i is : {0: 30,1: 00,1: 30, … … … … … … ..,24: 00} half hourly iteration.

Next, the coefficient of variation (CV) of the standard deviation for the 12 months was computed from the expression in equations (3.9) and (3.10). The values obtained were also recorded in Table 4.3.

$$CV_{MW} = \frac{\sigma_{P(i,j)}}{\bar{P}_{(i,j)}} \times 100\% \qquad\qquad 3.9$$

$$CV_{MVar} = \frac{\sigma_{Q(i,j)}}{\bar{Q}_{(i,j)}} \times 100\% \qquad\qquad 3.10$$

Similarly, the phase angle and power factor were computed accordingly after which the graphs of maximum active power and standard deviation of the maximum active power from the mean for the entire year were produced and presented in Table 4.3.

3.2 PROCEDURE FOR DETERMINING THE LOAD CURRENT LIMITATIONS OF THE OSOGBO-AKURE 132 KV TRANSMISSION LINE.

3.2.1 Determination of Daily Maximum and Minimum Active Load Values

The maximum active and reactive power components for the year were obtained from the 30 minutes power transmission detail of each day of the 12 months through a half-hourly statistical method of time series data. The MAX MW and MAX MVar values are displayed at the base of the second and third columns of Table 4.3. The algorithm is presented in equations (3.11) to (3.14).

$$P_{max(i)} = MAX \{P_{i,j}\} \hspace{3cm} 3.11$$

$$Q_{max(i)} = MAX \{Q_{i,j}\} \hspace{3cm} 3.12$$

$$P_{min(i)} = MIN \{P_{i,j}\} \hspace{3cm} 3.13$$

$$Q_{min(i)} = MIN \{Q_{i,j}\} \hspace{3cm} 3.14$$

$$where \; j = \{1,2,3, \dots \dots \dots \dots \dots \dots .365\}$$

$$i = \{0:30, 1:00, 1:30 \dots \dots \dots \dots .24:00\}$$

$P_{max(i)}$ describes the daily load curve for the line.

Equally, a graph of the maximum active and reactive power for each half-hour in the month, mean active power and standard deviation (standard error) of active power for every half-hour against time for the twelve month period was presented in Chart 4.13 to examine the trend in the power flow over the entire year.

3.2.2 Determination of the Operating Conditions and Limitations of Osogbo-Akure 132 kV Transmission Line

The Osogbo-Akure 132 kV transmission line was represented in the equivalent circuit diagram as shown in figure 3.4. The method of successive or sequential approximations was used to determine the operating conditions and limitations of the Osogbo-Akure 132 kV transmission line through the following procedure.

36

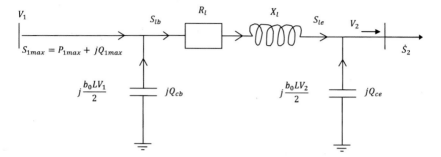

FIGURE 3.4 Equivalent (Impedance) Circuit of the Osogbo-Akure Power Transmission

3.2.2.1   Method of Successive Approximations.

Some of the network parameters of the Osogbo-Akure 132 kV radial line are enumerated below while the full compilation is contained in Table 4.1:

    i. r_0: $Per - km\ active\ resistance\ (\Omega/km)$;

    ii. x_0: $Per - km\ inductive\ reactance\ (\Omega/km)$;

    iii. b_0: $Per - km\ susceptance\ (S/km)$;

    iv. L: $Per - Length\ of\ line\ (km)$;

    v. N: $Number\ of\ circuits$;

    vi. $V_{nominal}$: $Nominal\ Voltage\ (km)$;

    vii. 1.05: $Nominal\ Voltage\ Constant$;

    viii. V_1: $V_{source} = Sending\ end\ voltage\ (kV) = V_{nominal} \times 1.05$;

    ix. V_2: $Recieving\ end\ voltage\ (kV)$.

These variables were used to compute the line parameters as presented in equations (3.15) to (3.18).

Line resistance (R_l) and reactance (X_l) are determined using equation 3.15 and 3.16.

$$R_l = \frac{r_0 \times L}{N}\ (\Omega): Line\ resistance \qquad\qquad 3.15$$

$$X_l = \frac{x_0 \times L}{N}\ (\Omega): Line\ reactance \qquad\qquad 3.16$$

$$Q_{cb} = \frac{b_0 \times L \times V_{source}^2}{2}\ (MVar): Line\ reactive\ power \qquad\qquad 3.17$$

$$Q_{ce} = \frac{b_0 \times L \times V_2}{2} \, (MVar) : Line \; susceptance \; at \; the \; end \; of \; line \qquad 3.18$$

The line mode for the transmission network was calculated in two approximation stages:

(1) First Approximation.

Here, the power distribution and losses in the network were determined. The initial step in computation of the power flow was the determination of the complex power (S_{mn}) components of the load at maximum and minimum mode operations as characterised in equation (3.19).

$$S_{mn} = (P_{mn} + jQ_{mn}) \; MVA \qquad 3.19$$

where m is point on the radial network;

n is maximum or minimum load flow mode;

S_n is apparent power component at maximum or minimum load flow mode;

P_n is active power component at maximum or minimum load flow mode;

Q_n is reactive power component at maximum or minimum load flow mode;

Having determined the maximum and minimum active components of the load flowing through the transmission line and the direction of power flow as characterized by figure 3.4, the mode parameters were computed as follows:

$$S_{lb} = P_{mn} + j(Q_{mn} + jQ_{cb}) \; MVA \qquad 3.20$$

$$S_{le} = (P_{lb} - \Delta P_l) + j(Q_{lb} - \Delta Q_l) \; MVA \qquad 3.21$$

where S_{lb} and S_{le} are values of apparent power

at the beginning and end of a section of the line

ΔP_l is active losses in the line; and ΔQ_l is reactive losses in the line

$$\Delta P_l = \frac{P_{lb}^2 + Q_{lb}^2}{V_1^2} R_l \; MW = Active \; component \; of \; losses \; on \; the \; line \qquad 3.22$$

$$\Delta Q_l = \frac{P_{lb}^2 + Q_{lb}^2}{V_1^2} X_l \; MVar = Reactive \; component \; of \; losses \; on \; the \; line \qquad 3.23$$

$$S_2 = P_2 + jQ_2 = P_{le} MW + j(Q_{le} + Q_{ce}) \; MVar \qquad 3.24$$

(2) Second Approximation

In this stage, the receiving end voltage was determined from the expression in equation (3.25) since the value of the sending end voltage was provided.

$$V_2 = \sqrt{\left(V_1 - \frac{P_{lb}R_l + Q_{lb}X_l}{V_1}\right)^2 + \left(\frac{P_{lb}X_l - Q_{lb}R_l}{V_1}\right)^2} \qquad\qquad 3.25$$

or

$$V_2 = \sqrt{(V_1 - \Delta V)^2 + (\delta V)^2} \qquad\qquad 3.26$$

where ΔV and δV are the direct and quadrature components
of voltage drop in the given section

Other steady state output mode parameters computed includes the following:

i. Percentage active and reactive component of the losses on the transmission line $\%\Delta P_l, \%\Delta Q_l$.

These are presented in the equations (3.27) and (3.28).

$$\%\Delta P_l = \frac{Active\ component\ of\ losses\ on\ the\ line}{Sending\ end\ maximum\ active\ power} \times 100$$

$$\%\Delta P_l = \frac{\Delta P_l}{P_{1max}} \times 100 \qquad\qquad 3.27$$

Similarly,

$$\%\Delta Q_l = \frac{Reactive\ component\ of\ losses\ on\ the\ line}{Sending\ end\ maximum\ reactive\ power} \times 100$$

$$\%\Delta Q_l = \frac{\Delta Q_l}{Q_{1max}} \times 100 \qquad\qquad 3.28$$

ii. Percentage voltage deviation $\%V_{dev}$.

The percentage voltage deviation of the line is described by equation (3.29).

$$\%V_{dev} = \frac{V_2 - V_{nominal}}{V_{nominal}} \times 100 \qquad\qquad 3.29$$

iii. Efficiency of the transmission line η .

This percentage efficiency of the transmission line was computed from equation (3.30).

$$\eta = \frac{Active\ power\ at\ the\ sending\ end}{Active\ power\ at\ the\ recieving\ end} \times 100$$

$$\eta = \frac{P_2}{P_{1max}} \times 100 \qquad\qquad 3.30$$

iv. Sending end apparent power $S_1(MVA)$.

The sending end apparent power was calculated from equation (3.31).

$$S_1(MVA) = \sqrt{P_{1max}^2 + Q_{1max}^2} \qquad\qquad 3.31$$

v. Receiving end apparent power $S_2(MVA)$.

The apparent power at the receiving end was computed from equation (3.32).

$$S_{2a}(MVA) = \sqrt{P_2^2 + Q_2^2} \qquad\qquad 3.32$$

An equivalent value of the receiving end apparent power was computed from equation (3.33).

$$S_{2b}(MVA) = \sqrt{3}IV_{nominal} \times 10^{-3} \qquad\qquad 3.33$$

$$where\ I = Thermal\ rating\ of\ the\ line\ in\ amperes$$

vi. The percentage loading on the transmission line $\%\ loading$.

The percentage loading on the transmission line in apparent power was calculated from equation (3.34).

$$\%loading = \frac{S_{2a}(MVA)}{S_{2b}(MVA)} \qquad\qquad 3.34$$

vii. The power factor of the sending end apparent power pf_1 .

The power factor of the sending end apparent power was computed from the equation (3.35).

$$pf_1 = cos_1 = \frac{P_{1max}}{S_1} \qquad\qquad 3.35$$

viii. The power factor of the receiving end apparent power pf_2 .

The power factor of the receiving end apparent power was computed from the equation (3.36).

$$pf_2 = cos_2 = \frac{P_2}{S_2} \qquad\qquad 3.36$$

Results obtained from equations (3.1) to (3.36) were tabulated accordingly in the Table 4.1.

3.2.3 Reliability Parameters Computation

3.2.3.1 Extraction of frequency and time duration.

Hourly load demand on the Osogbo-Akure 132 kV single circuit line and the annual average customer population data were obtained from the receiving end of the transmission line for a period of 60 months-i.e. 2005 to 2009. The frequency and time durations of electric power outages were extracted and compiled in Appendices B.1 to B.5. The electric power data was used to compute the failure rate. This is because the failure rate of a system is usually dependent on time, with the rate varying over the life cycle of the system.

3.2.3.2 Failure rate λ .

This parameter was computed for each year from the equation (3.37).

$$\lambda = \frac{Total\ number\ of\ failures}{Number\ of\ days\ in\ a\ year} \qquad 3.37$$

3.2.3.3 The Mean Time Between Failure (MTBF).

The MTBF is another important reliability parameter that was calculated for each of the details of the five years under consideration. This is described by equation (3.38).

$$MTBF = Mean\ uptime = \frac{\sum_{i=1}^{365} \Delta t_{up(i)}}{Number\ of\ failures} \qquad 3.38$$

where $\Delta t_{up(i)}$ is the ith nominal operating time

The values of every reliability parameter obtained for each of the years were computed by extracting details from appendices A.1 to A.5 and B1 to B5. These results are compiled in TABLE 4.4.

3.2.3.4 Customer-oriented Interruption Indices.

In addition to the determination of the failure rate, the customer-oriented indices for each year, which are also important reliability parameters, were also computed from the expressions in equations

System average interruption frequency index, SAIFI

$$SAIFI = \frac{Total\ number\ of\ customer\_interruptions}{Total\ number\ of\ customers\ served} = \frac{\Sigma \lambda_i N_i}{\Sigma N_i} \qquad 3.39$$

where λ_i is the failure rate and N_i is the number of customers at load point i

41

Customer average interruption frequency index, CAIFI

$$CAIFI = \frac{Total\ number\ of\ customer\_interruptions}{Total\ number\ of\ customers\ affected} = \frac{\sum \lambda_i N_i}{\sum_j Na_j} \qquad 3.40$$

System average interruption duration index, SAIDI

$$SAIDI = \frac{Sum\ of\ customer\_interruption\ durations}{Total\ number\ of\ customers} = \frac{\Sigma U_i N_i}{\Sigma N_i} \qquad 3.41$$

where U_i is the annual outage time and N_i is the number of customers of load point i

Customer average interruption duration index, CAIDI

$$CAIDI = \frac{Sum\ of\ customer\_interruption\ durations}{Total\ number\ of\ customer\ interruptions} = \frac{\Sigma U_i N_i}{\Sigma \lambda_i} \qquad 3.42$$

where λ_i is the failure rate, U_i is the annual outage time

and N_i is the number of customers of load point i

Average service availability index, ASAI

$$ASAI = \frac{Customer\_hours\ of\ available\ service}{Customer\ hours\ demanded} = \frac{\Sigma U_i \times 341 - \Sigma U_i N_i}{\Sigma N_i \times 341} = \qquad 3.43$$

Average service unavailability index, ASUI

$$ASUI = 1 - ASAI = \frac{Customer\_hours\ of\ unavailable\ service}{Customer\ hours\ demanded} = \frac{\Sigma U_i N_i}{\Sigma N_i \times 341} = \quad 3.44$$

3.3 EVALUATION OF THE POSSIBILITY OF FEEDING AKURE FROM ALTERNATE SOURCE

3.3.1 Consideration of Viable Electric Power Sources

A number of alternative sources of electric power supply were considered for Akure district distribution network. These options are:

a) Osogbo-Benin 330 kV transmission line.

b) A separate Osogbo-Akure 132 kV transmission line.

c) Omotosho power plant.

d) Owena Dam.

3.3.2 Two Electric Power Sources Closely Considered

Options a) and b) were selected out of the four options initially considered for close examination. The reason for this choice is because these transmission lines have existing sources. Options c) and d) depend on primary variables, without which they cannot be potential sources of electric power supply. For this reason, these first two options were examined based on the following factors:

1. Proximity to Akure
2. Present load capacity

3.3.3 Information on the Components Required for Execution of the Selected Electric Power Sources.

3.3.3.1 Tee-off from the Osogbo-Benin 330 kV transmission line

Osogbo-Benin 330 kV transmission line passes through Akure town and is approximately 95 meters from the existing 132 kV substation. Summary of the present average load supplied to Benin 330 kV switching substation every month is 152,552,000 kWh while the average amount of electric energy it dispatches is 83,142,000 kWh. This implies that it has a surplus of 69,410,000 kWh (PHCN, 2005). This quantity of electrical energy is about four times the average monthly load requirement at the Akure distribution network. This is illustrated in appendix B. The cost and items required to tee-off from Osogbo-Benin transmission line and run a short length connection to the existing substation is listed in Table 3.2. This can be executed within a period of 6 months. A similar connection was executed in Aba in 2008 within this period (PHCN, 2005).

3.3.3.2 A Separate Osogbo-Akure 132 kV Transmission Line

Another viable option of an alternative source of electric power supply to Akure is another 132 kV transmission line erected from Osogbo. Its source is about 93 km from Akure. There are two transformers currently serving Akure feeder as can be observed in Figure 3.1. These are transformers 4T6-150 MVA and 4T2-90 MVA with loading capacities of 97.08% and 96.67% respectively. These values are obtained from Table 4.16. The transformer 4T1-150 MVA was recently replaced after an explosion. It is presently serving as redundancy. The three transformers can be engaged in parallel for the purpose of sustaining an increased load demand on Akure district distribution system through another 132 kV transmission line. The cost and items required to install another Osogbo-Akure 132 kV transmission line is listed in Table 3.3. This option can be executed within 20 months.

43

TABLE: 4.16: Maximum Load Reading Log for 2010.

S/N	WORK CENTRES	SUB-STATION	TRFS. NOM. & RATING	AVAILABLE CAPACITY (MVA)	VOLTAGE RATING	MAKE	MAXIMUM LOAD FOR THE MONTH OF JUNE, 2010.								MAX. LOAD TO DATE		
							MW	MVA	AMP	MVA	% LOADING	TIME	DATE	WINDING TEMP °C	MW	TIME	DATE
1	OSOGBO AREA CONTROL CENTRE	OSOGBO 330KV	4T1-150MVA	150	330/132/33	PAUWELS	BURNT. REPLACEMENT IN PROGRESS										
2			4T6-150MVA	150	330/132/33	AREVA	116.5	-	240	145.6	97.08	14:00	18/06 /10	64	106.8	22:0 0	16/12/ 09
3			4T2-90MVA	90	330/132/13.8	MITSUBISHI	69.6	-	148	87.0	96.67	06:00	20/06 /10	62	62.2	20:3 0	28/05/ 09
4		OSOGBO 132KV	4T3-60MVA	60	132/33	T & R	42.3		846	52.9	88.13	20:00	21/06 /10	60	48.0	20:3 0	29/10/ 09
5			4T4-60MVA	60	132/33	T & R	44.1		882	55.1	91.88	17:00	05/06 /10	58	44.1	17:0 0	05/06/ 10
6			4T5-30MVA	30	132/33	MITSUBISHI	19.5	-	390	24.4	81.25	19:00	16/06 /10	61	21.5	16:3 0	05/06/ 09
7		IFE	T1-30MVA	30	132/33	ELTA	12.0	6.0	240	15.0	50.00	09:00	29/06 /10	47	14.8	20:0 0	12/03/ 10
8			T2-30MVA	30	132/33	ELTA	12.0	6.0	240	15.0	50.00	09:00	29/06 /10	47	14.8	20:0 0	12/03/ 10
9		ILESHA	T1-40/50MVA	40	132/33	ABB	23.8		476	29.8	74.38	06:00	30/06 /10	57	25.5	19:0 0	01/05/ 09
10			T2-40/50MVA	40	132/33	ABB	8.4	-	168	10.5	26.25	20:00	08/06 /10	50	19.7	18:0 0	22/06/ 09
11		OFFA	T1-30MVA	30	132/33	TOSHIBA	13.6	-	272	17.0	56.67	21:00	18/06 /10	46	14.9	06:0 0	01/12/ 08

SOURCE: Archives of National Control Centre, Osogbo, Osun State Nigeria.

3.3.4   Design and Sizing of Proposed Substation Circuit Elements

3.3.4.1   Transformer Capacity Design and Sizing

The proposed 330/132 kV substation transformer design specifications observed are:

a) Two transformers shall be installed in parallel to service the substation.

b) Each of these transformers shall carry half of the designed substation connected load current at normal steady-state operating mode.

c) Each transformer shall be capable of carrying the load demand of the demand centre in case of the failure of one.

The transformer capacity S_T (kVA) was determined from the expression in equation (3.45) where the values of $P_{max\,2}$ and $Q_{max\,2}$ were obtained from Table 4.1.

$$S_{max\,2} = \sqrt{P_{max\,2}^2 + Q_{max\,2}^2} = S_2 \; MVA \qquad\qquad 3.45$$

$$S_{max\,2} = \sqrt{69.86^2 + 38.58^2} = 79.80 \; MVA$$

where $S_{max\,2}$ = $Maximum\ apparent\ power\ transmitted$

$P_{max\,2}$ = $Maximum\ active\ power\ component\ transmitted$

$Q_{max\,2}$ = $Maximum\ reactive\ power\ component\ transmitted$

The proposed transformer selection was made from the design capacity computation. The transformer capacity was selected from the relationship in equation (3.46).

$$S_{max\,2} \leq 1.2 S_T \qquad\qquad 3.46$$

$$S_T \geq \frac{S_{max\,2}}{1.2}, MVA$$

$$S_T \geq \frac{79.80}{1.2} = 66.50 \; MVA \qquad\qquad 3.47$$

where = S_T $is\ the\ number\ of\ transformers$

$1.2 = permissible\ overload\ coefficient$

The expression in equation (3.47) shows that the transformer capacity should be greater or equal to 66.50 MVA. No standard transformer with the capacity of the value obtained from the application of equation (3.47) exists. Therefore, two 60 MVA capacity transformers were selected.

The transformer loading coefficient LC_T for the two transformers was computed from the terms in equation (3.48).

$$LC_T = \frac{S_{\max 2}}{n \times S_T} \qquad \qquad 3.48$$

$$LC_T = \frac{79.80 \ MVA}{2 \times 66.50} = 0.60 \qquad \qquad 3.49$$

$where \ S_{\max 2} = Maximum \ apparent \ power \ transmitted$

$n = Number \ of \ transformers \ paralleled \ at \ 120\% \ load$

The result obtained from equation (3.49) implies that the two transformers are load by 60%. In other words, 40% of the total capacity of each of the two transformers will be available in event of an increased load demand in Akure metropolis.

3.3.4.2   Sizing of Fuse and Circuit Breaker

A circuit breaker ampacity factor is 2.5 (maximum permissible current) of the circuit or apparatus it is installed to protect. Similarly, the fuse is 1.75 (maximum permissible current) for time delay (NEC 430-152, 2008). Lower multiplication factor was adopted to allow the breaker or fuse to be closer to starting and operating characteristics, thereby allowing the over current device to trip or melt at lower levels of current. Fuses and circuits breakers were selected based on the following criteria:

a) The nominal current permissible under normal working condition
b) The amount of current to be interrupted

The circuit breaking capacity was determined by equation 3.50

$$I_N = \frac{S}{\sqrt{3} \times U_N} \qquad \qquad 3.50$$

$$I_N = \frac{60 \times 10^6}{\sqrt{3} \times 330 \times 10^3} = 3.499 \ A$$

$where \ I_N = Nominal \ current$

$S = Apparent \ power \ capacity \ of \ transformer \ selected$

$U_N = Nominal \ Voltage$

3.3.4.3   Isolator Sizing.

Circuit breaker and isolators are installed in series. Hence, both are expected to permit the same magnitude of current under normal and abnormal conditions. For this reason, the same capacity of circuit breaker was selected.

3.3.5   Design of Proposed 330/132 kV Substation

The proposed 330/132 kV substation was designed as illustrated in Fig 3.5.

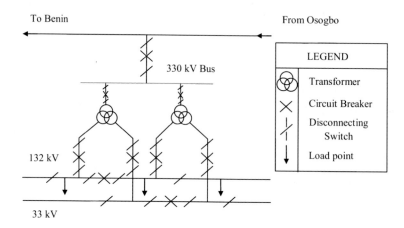

FIGURE 3.5 Design of Proposed 330/132 kV Substation

3.3.6   Preparation of the Bill of Engineering Measurement and Evaluation (BEME)

Bills of Engineering Measurement and Evaluation were prepared for each of the two options considered as alternative electric power supply sources for Akure district distribution network.

TABLE 3.2.Bill of Engineering Measurement and Evaluation (BEME) for proposed tee-off from the existing 330/132 kV transmission line

Item	Substation Equipment	Quantity	Spec/Brand	Unit Cost ₦	Cumm. Cost ₦
1	Transformer	2	150 MVA 330/132 kV Siemens	400,000,000	800,000,000

47

2	Reactor	1	75 MVA	2,845,000	2,845,000
3	Interposing Voltage Transformer	3	330 kV	1,200,000	3,600,000
4	Current Transformer	12	Ratio:1500/750/1A, 1600/800/1A	2,845,000	34,140,000
5	Circuit Breaker	6	330 kV	24,662,400	147,974,400
6	Isolators with earth switch	5 sets	330 kV	4,024,613	20,123,065
7	Control/relay panel	1	330 kV	4,672,500	4,672,500
8	Lead Acid Battery Bank	3 banks	110V D.C 400AH	3,453,206	10,359,618
9	3-phase Battery Charger	1	110V D.C 415V 3-phase	1,583,255	1,583,255
10	Capacitor Voltage Transformer (CVT)	1	330 kV	1,629,500	1,629,500
11	330 kV Lightning Arrester	6	331 kV	982,000	5,892,000
12	Base radio (V.H.F)	1	Very high frequency (V.H.F)	2,200,000	2,200,000
13	Transceivers	5		80,000	400,000
14	H.F Radio	2		2,000,000	4,000,000
15	Pabx	3		5,000,000	15,000,000
16	Synchronizing Check Relay	2	330 kV ANSI 25, Case 4	915,200	1,830,400
17	Stand-by Generator	1	27 kVA 3-phase	2,096,788	2,096,788
18	Aluminium Conductor	1 km	350mm Mic-Com Cable	2,350,000	2,350,000
				TOTAL	1,060,696,526

TABLE 3.3. Bill of Engineering Measurement and Evaluation for Proposed 132/33 kV Transmission Line and Substation Equipment.

Item	Substation Equipment	Quantity	Spec/Brand	Unit Cost-₦	Cumm. Cost ₦
1	Transformer	2	60 MVA 132/33 kV Siemens	260,000,000	520,000,000
2	Reactor	1	30MVA	1,433,265	1,433,265
3	Interposing Voltage Transformer	3	132 kV	960,000	2,880,000
4	Current Transformer	12	132 kV	1,433,265	17,199,180
5	Circuit Breaker	6	133 kV	6,382,080	38,292,480
6	Isolators and earth switch	5 sets	132 kV Pantograph/centre break	1,600,000	8,000,000
7	Control/relay panel	1	132 kV	2,670,000	2,670,000
8	Battery Bank	1 bank	48/50V D.C 200AH	1,100,000	1,100,000
9	3-phase Battery Charger	1	48/50V D.C 3-phase	950,000	950,000
10	Capacitor Voltage Transformer (CVT)	1	132 kV	998,950	998,950
11	132 kV Lightning Arrester	6	132 kV	380,000	2,280,000
12	Base radio (V.H.F)	1	Very high frequency (V.H.F)	2,200,000	2,200,000
13	Transceivers	3		80,000	240,000
14	H.F Radio	2		2,000,000	4,000,000
15	Pabx	3		5,000,000	15,000,000
12	Transformer Differential Relay	2	VAJH 31 DFA 4C	645,000	1,290,000
13	Steel Towers and Accessories	366	132 kV	3,950,450	1,445,864,700

14	93 km by 60 metres Right-of-Way	2046 plots	93 km by 60 metres	250,000	511,500,000
15	250mm Aluminium Conductor	94 km		1,762,500	165,675,000
				TOTAL	2,741,573,575

CHAPTER FOUR

RESULTS AND ANALYSIS

4.1 RESULTS

4.1.1 Voltage and Efficiency Profile.

TABLE 4.1 VOLTAGE AND EFFICIENCY PROFILE

Steady State Load Flow Mode			
INPUT			
Network Variable	Symbol	Value	
		max mode	min mode
Per-km active resistance (Ω/km)	r_0	0.2733	0.2733
Per-km inductive reactance (Ω/km)	x_0	0.4445	0.4445
Per-km Susceptance (Sm/km)	b_0	2.51E-06	2.51E-06
Length of line (km)	L	93	93
Number of circuits	N	1	1
Nominal Voltage (kV)	$V_{nominal}$	132	132
Nominal Voltage Constant	1.05	1.05	1.05
Conductor Size (mm²)	250	250	250
Thermal Rating (Amps)	550	550	550
Surge Impedance (Ω)	417	417	417
Mode Parameters			
V_1 (kV)	V_{source}	138.6	138.6
P_{Le} (MW)	P_{Le}	69.86	48.68
Q_{Le} (MVar)	Q_{Le}	37.25	24.59
P_{Lb} (MW)	P_{Lb}	84.00	54.00
Q_{Lb}(MVar)	Q_{Lb}	60.24	33.24
Q_{cb} (MVar)	Q_{cb}	2.24	2.24
Q_{ce} (MVar)	Q_{ce}	1.33	1.67
P_{1max}(MW)	P_{1max}	84	54
Q_{1max}(MVar)	Q_{1max}	58	31
OUTPUT			
V_2 (kV)	V_2	106.87	119.91
ΔP_L (MW)	ΔP_L	14.14	5.32
ΔQ_L (MVar)	ΔQ_L	22.99	8.65
ΔP_L (MW)%	ΔP_L%	16.83	9.85
ΔQ_L (MVar)%	ΔQ_L%	39.64	27.91
P_2 (MW)	P_2	69.86	48.68
Q_2(MVar)	Q_2	38.58	26.26
$V_{deviation}$ %	V_{dev}	-19.03	-9.16
Efficiency %	η	83.2	90.1
S_1 (MVA)	S_1	102.08	62.27
S_{2a} (MVA)	S_{2a}	79.81	55.31
S_{2b} (MVA)	S_{2b}	125.75	
% Loading S_{2a}/S_{2b}	S_{2a}/S_{2b}	63	
pf$_1$	$\cos\theta_1$	0.82	0.87
pf$_2$	$\cos\theta_2$	0.88	0.88

TABLE 4.2 Half hourly Statistics of P (MW) and Q (MVar) for 2009.

Minut es	MAX, MW	MAX, MVar	MEA N, (MW)	MEA N, (MVa r)	STDV A, (MW)	Coeff of Var	STDV A, (MVa r)	Coeff of Var	Phase angle of MVA (θ)	Power factor
0	68	32	58.73	30.67	4.80	8.17	1.44	4.68	25.20	0.90
00:30	58	48	56.42	35.00	1.62	2.87	8.00	22.86	39.61	0.77
01:00	70	32	56.33	29.83	5.10	9.06	2.08	6.98	24.57	0.91
01:30	57	32	55.27	29.42	1.68	3.04	2.61	8.87	29.31	0.87
02:00	58	31	55.50	29.42	2.54	4.58	1.98	6.71	28.12	0.88
02:30	58	31	55.50	29.67	2.61	4.70	1.30	4.39	28.12	0.88
03:00	57	31	54.08	29.00	2.19	4.06	2.17	7.50	28.54	0.88
03:30	57	31	54.75	29.08	1.76	3.22	1.98	6.79	28.54	0.88
04:00	57	35	53.75	29.00	2.53	4.70	2.59	8.94	31.55	0.85
04:30	57	35	53.92	29.00	2.61	4.84	2.56	8.82	31.55	0.85
05:00	57	36	55.25	29.75	2.18	3.94	2.60	8.73	32.28	0.85
05:30	58	37	55.75	32.67	1.60	2.87	3.06	9.35	32.54	0.84
06:00	61	47	58.08	35.08	2.27	3.92	5.07	14.46	37.61	0.79
06:30	65	35	59.67	32.58	3.14	5.27	1.73	5.31	28.30	0.88
07:00	63	36	59.92	33.58	2.91	4.85	2.27	6.77	29.74	0.87
07:30	68	36	60.08	33.58	4.38	7.29	2.15	6.41	27.90	0.88
08:00	66	42	59.75	34.50	3.57	5.98	3.99	11.56	32.47	0.84
08:30	70	38	62.25	34.50	5.85	9.40	2.94	8.52	28.50	0.88
09:00	79	38	61.33	34.58	7.75	12.64	2.91	8.40	25.69	0.90
09:30	60	36	56.67	33.25	3.17	5.60	2.22	6.68	30.96	0.86
10:00	60	37	56.00	33.92	3.16	5.65	2.39	7.05	31.66	0.85
10:30	60	38	55.75	34.00	3.25	5.83	2.98	8.78	32.35	0.84
11:00	55	37	53.42	33.42	1.62	3.04	2.71	8.12	33.93	0.83
11:30	56	34	54.33	32.50	1.61	2.97	1.31	4.04	31.26	0.85
12:00	56	36	54.00	32.50	2.19	4.06	1.88	5.79	32.74	0.84
12:30	54	34	52.00	30.92	1.81	3.48	1.16	3.77	32.20	0.85
13:00	56	34	54.08	32.75	1.78	3.29	1.42	4.34	31.26	0.85
13:30	57	39	53.83	34.17	2.72	5.06	3.79	11.08	34.38	0.83
14:00	56	34	53.67	31.75	2.84	5.29	1.91	6.02	31.26	0.85
14:30	56	35	53.75	32.17	2.77	5.15	2.29	7.12	32.01	0.85
15:00	56	34	54.33	31.83	2.19	4.03	1.70	5.33	31.26	0.85
15:30	56	34	54.00	32.08	2.09	3.87	2.07	6.44	31.26	0.85
16:00	58	35	54.42	32.25	2.27	4.18	2.14	6.63	31.11	0.86
16:30	62	35	55.42	31.83	4.01	7.24	3.38	10.62	29.45	0.87

17:00	59	38	54.00	32.42	3.77	6.97	3.58	11.04	32.78	0.84
17:30	68	38	57.58	32.17	7.73	13.42	3.97	12.35	29.20	0.87
18:00	68	38	57.83	32.58	6.32	10.93	3.20	9.83	29.20	0.87
18:30	64	41	59.42	35.08	3.12	5.25	3.09	8.80	32.64	0.84
19:00	66	41	61.83	37.08	2.29	3.70	2.23	6.03	31.85	0.85
19:30	66	40	61.75	37.92	2.99	4.84	1.73	4.56	31.22	0.86
20:00	84	58	68.64	42.75	10.02	14.60	9.27	21.70	34.62	0.82
20:30	66	50	61.82	39.00	2.56	4.14	6.70	17.18	37.15	0.80
21:00	65	38	60.25	35.83	1.82	3.01	2.12	5.93	30.31	0.86
21:30	62	37	60.50	34.67	1.45	2.39	1.61	4.66	30.83	0.86
22:00	62	38	60.58	35.67	1.62	2.68	2.19	6.13	31.50	0.85
22:30	63	38	61.50	35.50	1.98	3.21	2.11	5.95	31.10	0.86
23:00	61	50	58.83	38.50	1.59	2.70	7.38	19.17	39.34	0.77
23:30	60	33	57.83	32.17	1.80	3.11	1.34	4.16	28.81	0.88
24:00:00	60	32	57.50	30.83	2.02	3.52	1.40	4.55	28.07	0.88
	84.00	**58.00**					**1.16**			
	54.00	**31.00**								

4.1.2 Reliability Assessment

TABLE 4.3: Reliability Assessment.

Reliability Assessment						
Customer Oriented Interruption Indices	Symbol	Value				
		2005	2006	2007	2008	2009
Number of outage incidences		89	108	246	184	194
Outage duration, (hrs)		126.98	180.97	536.78	356.93	274.87
Failure rate (failures/year)	λ_i	89.00	108.00	246.00	184.00	194.00
Mean Time Between Failure (hrs/failure)	MTBF	97.00	79.44	33.43	45.67	43.74
Number of customers on load point i	N_i	47677	47729	47806	47919	48054
Number of load points	i	1	1	1	1	1
Annual outage time (hrs)	U_i	127	181	537	357	275
Active customer population		44538	44576	44629	44698	44806
Inactive customer population		3139	3153	3177	3221	3248
Total number of customers	$\sum N_i$	47677	47729	47806	47919	48054
Total number of customer-interruptions	$\sum \lambda_i N_i$	4243253	5154732	11760276	8817096	9322476
Sum of customer-interruption durations, customer (hrs)	$\sum U_i N_i$	6054979	8638949	25671822	17107083	13214850
Customer hours of available service		8633	8579	8223	8403	8485
Customer hours of unavailable service		127	181	537	357	275
Customer hours demanded		8760	8760	8760	8760	8760
System Average Interruption Frequency Index	SAIFI	89.00	108.00	246.00	184.00	194.00
System Average Interruption Duration Index (hrs)	SAIDI	126.98	180.97	536.78	356.93	274.87
Customer Average Interruption Frequency Index	CAIFI	89	108	246	184	194
Customer Average Interruption Duration Index (hrs)	CAIDI	1.43	1.68	2.18	1.94	1.42
Average Service Availability Index	ASAI	0.99	0.98	0.94	0.96	0.97
Average Service Unavailability Index	ASUI	0.01	0.02	0.06	0.04	0.03

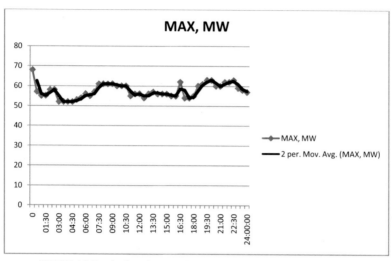

CHART 4.1 Graph of maximum active power against time for the month of January,
2009

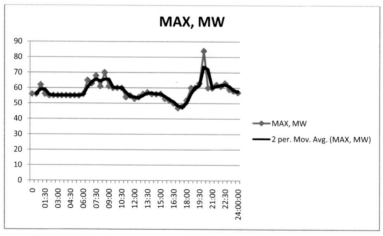

CHART 4.2 Graph of maximum active power against time for the month of February,
2009

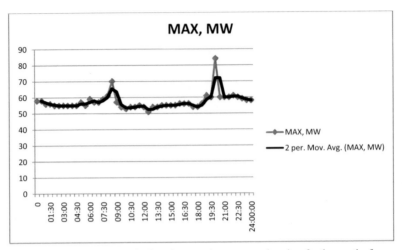

CHART 4.3 Graph of maximum active power against time for the month of
March, 2009

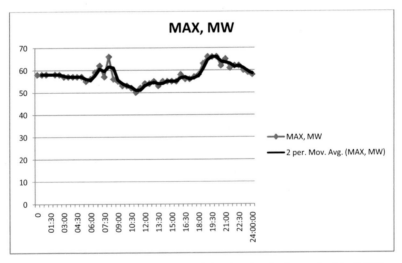

CHART 4.4 Graphs of maximum active against time for the month of April, 2009

56

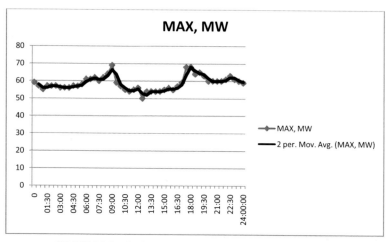

CHART 4.5 Graph of maximum active power against time for the month of
May, 2009

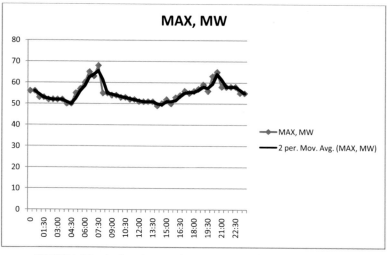

CHART 4.6 Graph of maximum active power and against time for the month of June,
2009

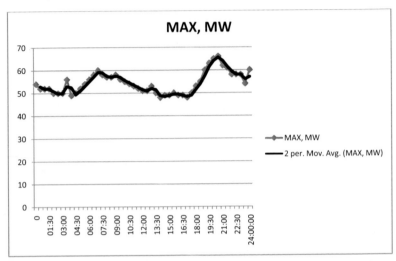

CHART 4.7 Graph of maximum active power against time for the month of

July, 2009

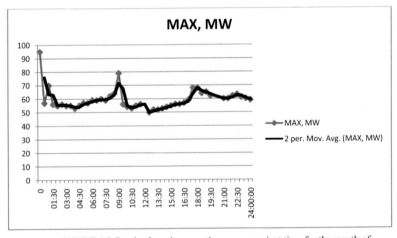

CHART 4.8 Graph of maximum active power against time for the month of

August, 2009

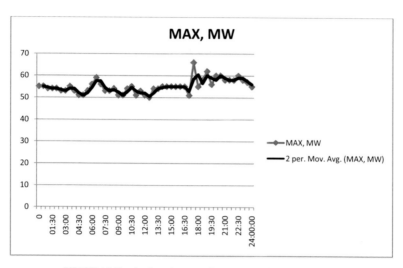

CHART 4.9 Graph of maximum active power against time for the month of
September, 2009

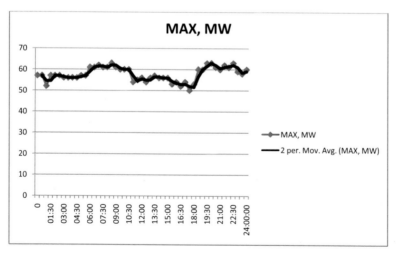

CHART 4.10 Graph of maximum active power against time for the month of October,
2009

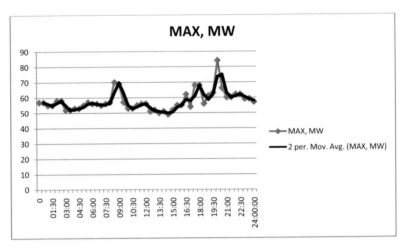

CHART 4.11 Graph of maximum active power against time for the month of
November, 2009

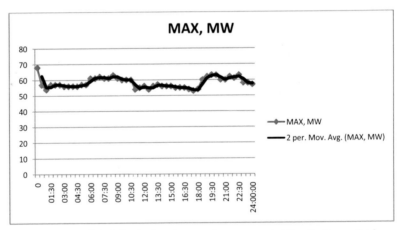

CHART 4.12 Graph of maximum active power against time for the month of
December, 2009

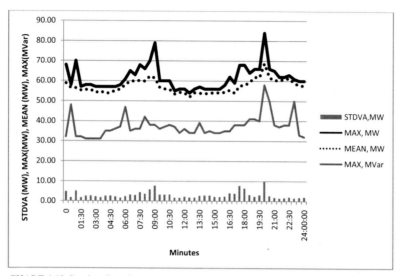

CHART 4.13 Graphs of maximum active and reactive power for each half-hour in the month, mean active power and standard deviation (standard error) of active power for every half-hour against time for the 2009

4.2. ANALYSIS

4.2.1 Statistical Estimation of the Daily Load Curve

Daily load curve was plotted from the half-hourly active and reactive power components (MW and MVar) obtained for a period of 12 months from the National Control Centre, Osogbo in order to examine the irregularities in the power transmission every 24 hours. A sample can be observed in figure 3.3. Consequently, the daily details were filtered and the load curve was plotted for each of the months as presented in Chart 4.1 to 4.12. The power components reached their peak values between 7:30-9:30am and 7:45-9:40pm monthly. These periods represent the monthly peak periods. The purpose of the Chart is to observe the consistency and variation in the power transmission profile for each of the 12 months.

4.2.2. Determination of the Current Limitations of the Osogbo-Akure 132 kV Transmission Line

The active and reactive components of power losses computed for the line were very high. 14.14 MW, 22.99 MVar and 5.32 MW, 8.65 MVar were obtained for ΔP_l and ΔQ_l at maximum and minimum operating conditions respectively. These values are very significant and cannot be attributed to technical losses alone. An inference is that there are few distribution transformers at the Akure end of the line. Fig 3.2 presents 10 distribution transformers of different capacities. All of these transformers combined are insufficient for effective distribution of 54 MW- which is the maximum active power component transferred at minimum load flow mode, let alone the 84 MW maximum active power component transferred at maximum load flow mode. Another deduction is the irregularity in the load allocation from the Osogbo end of the transmission line. The maximum active power transferred-84 MW-occurred once in the entire data for 2009. This might have been allocated or recorded in error as the value never repeated itself in the log compiled for the year.

The single circuit radial line has a percentage voltage deviation of -19.03% and -9.16% at maximum and minimum mode operations. These values compiled in Table 4.1 fall within ±5% which is the recommended permissible maximum according to the Institute of Electrical Engineers (IEE), U.K regulation. The transmission line has efficiencies of 83.2% and 90.1% at maximum and minimum mode operations respectively. These values presented in Table 4.1 are very high and reflective of the extent of support the Osogbo-

Akure 132 kV line is offering to the process of electric power transmission to Akure metropolis and the south-western part of Nigeria.

Table 4.1 presents 63% as the percentage loading on the transmission line within the twelve months considered. Within a short-term consideration, the transmission line can conveniently support an additional load of 37% capacity in event of an increase in electric power demand in Akure metropolis.

Maximum and mean half hourly estimates of active and reactive power: The half-hourly active and reactive power readings obtained from the sending end of the Osogbo-Akure 132 kV radial line were filtered through a Half-hourly Time Series Statistical Analysis Procedure. The first reason was to identify any outrageous value recorded in the process of typing in the values at the Osogbo end of the transmission line. An example is active power recorded at 12:00 minutes, 20:00 minutes and 20:30 minutes in Appendix A8. The second intent was to accurately identify the maximum active and reactive power components transmitted through the line at every half hour of each month. This is because the most common way utilities quantify a circuit's load is the peak demand over a specific period of time. The third and final reason was to establish the maximum and minimum MAX, MW and MAX, MVar for the whole year.

These values are presented in the second and third columns of Table 4.3 under the captions MAX MW and MAX MVar respectively. Subsequently, the mean active and reactive power components (MEAN MW and MEAN MVar) were determined and compiled in the same table. Afterwards, the extent of deviation of the maximum active and reactive power components from their estimated mean values for 30 minutes duration was determined by computing their standard deviations. The results obtained for each of the 12 months were recorded as STDVA, MW and STDVA, MVar. Similarly the coefficient of variation of the active and reactive component of standard deviation was estimated for the 12 months duration. The values obtained were tabulated in Table 4.3. This index depicts the consistency of active and reactive power components. The Half-hourly Time Series Statistical Procedure was repeated for every 30 minutes detail of each month in order to arrive at a compilation for the year. This is presented in Table 4.3. The maximum MAX, MW and MAX, MVar for the whole year was identified as 84 MW and 58 MVar. 54 MW and 31 MVar were identified as minimum power constituents. These values served as the initial active and reactive components of the sending end apparent power at maximum and

63

minimum mode operations as characterised in equation 3.5. The peak and the least values of active and reactive power were utilized for the assessment of the operation on the line. This is with a view to determining whether the line can support electric power transmission at maximum and minimum operating conditions.

The most useful of the component of electric power transferred is the active power. This constituent of the apparent power transferred is what is actually utilized at the various load points. Apparently, electrical load of a customer is the sum of the load drawn by the customer's individual appliances. Therefore, electricity tariff is charged on every kilowatt of active power utilized at every hour. Values in the STDVA, MW column of Appendices A1 to A12, that exceed 10 are very significant deviations of the MAX, MW from the corresponding mean active power values. However, STDVA, MW values less than 10 fall within permissible limits. Their graphs were plotted in Charts 4.1 to 4.12 to clearly illustrate the monthly variation. Chart 4.13 is a compilation of the graphs of the STDVA MW, MAX MW, MEAN MW, and MAX MVar for twelve months. It particularly presents the disparity between the maximum and mean active power.

The phase angle of the apparent power supplied at every half hour was computed and their corresponding power factor values were calculated. The values obtained for the entire year are compiled in Table 4.3. Power factor values that are less than 0.8 indicate a widely varying electric power supply. Significant voltage fluctuations do not support the smooth operation of some electrical equipment. Examples of such appliances are the electrocardiograph (a life support equipment) and some testing appliances being used in the various technical laboratories in the Akure. However, power factor values greater than 0.8 indicates that the load runs almost constantly. This is preferable from the utility point of view. It helps most electrical appliances in the south western metropolis to work long with high technical and economic indices.

4.2.3. Reliability Assessment

The regularity with which an engineered system falls short or the total number of failures within an item population is an important reliability index. The frequency and corresponding duration of electric power outage for 2005 to 2009 was extracted from the yearly load demand data. These were compiled and tabulated in appendices C.1 to C.5. The summation of the total time duration for all outage occurrences in each of these years corresponding to its failure rate were recorded in Table 4.4.

TABLE 4.4: Outage Records

Year	Outage duration, (hrs)	Cumulative, (hrs)	Mean outage duration, (days/yr)	Mean outage duration, (mins/day)
2005	126.98	126.98	5.0	20
2006	180.97	307.95	7.0	29
2007	536.78	717.75	22.0	88
2008	356.93	893.72	14.0	58
2009	274.87	631.80	11.0	45

The graph of the outage duration in days against years was also plotted as illustrated in Chart 4.14.

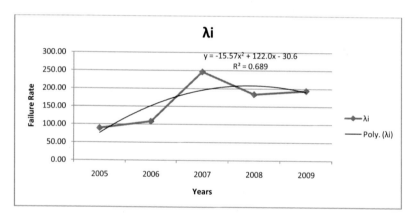

CHART 4.15 Graph of failure rates against years

These values were plotted against the total number of days in a year and presented in Chart 4.15. The highest failure rate occurred in 2007. This corresponds to mean outage duration of 88 minutes per day as illustrated in Table 4.4. A second order polynomial trend line was drawn across the graph of the yearly failure rates with corresponding regression equation. The equation of the line was deduced as presented in equation (4.1).

65

$$y = -15.57x^2 + 122x + 30.6 \qquad\qquad 4.1$$

Equation (4.1) is useful in predicting future outage patterns on the Osogbo-Akure 132 kV radial line. Their projective manifestations are important factors in the insurance, business and regulation practices as well as fundamental to design of safe systems. Forecast can be achieved by comparing the equation with a typical second order polynomial equation (4.2).

$$y = \alpha x^2 + \beta x + c \qquad\qquad 4.2$$

$where\ y\ is\ Failure\ rate$

$x\ is\ projected\ year$

$c\ is\ point\ of\ interception\ of\ trendline\ with\ vertical\ axis$

Afterwards, projective failure rate y can be computed by substituting that year for x in equation (4.1). This is because the point of interception $c = 30.6$ of the curve is constant.

Similarly, the amount of kilowatt of energy lost for each year can be computed from the expression in equation (4.3).

$$Tariff\ for\ 1\ kilowatt\ of\ power \times failure\ rate \qquad\qquad 4.3$$

For 2009, the amount of kilowatt of energy (kWh) lost is calculated as contained in equation (4.4).

$$5.90 \times 194 = 1144.6\ \text{kWh} \qquad\qquad 4.4$$

Mean Time between Failure (MTBF): This is an important reliability parameter specifically useful system where failure rate needs to be managed. The MTBF was computed for each year from 2005 to 2009 and presented in Table 4.3 with the highest value of 0.0112 occurring in 2005. All the values of MTBFs calculated fall below one. This trend is an indication of the effectiveness of the response of the maintenance crew of the PHCN to outage occurrences on that transmission line. MTBF is a parameter that appears frequently in engineering design requirements, and governs frequency of required system maintenance and inspections.

Customer-Oriented Indices: Average customer population data was obtained for each year from 2005 to 2009 and compiled in appendix C. These details were used to estimate the customer oriented indices compiled in Table 4.3. These parameters are very useful for

66

assessing the severity or significance of system failures in future reliability prediction analysis. They can also be used as a means of assessing the past performance of the system.

4.2.4. Appraisal of the Options Considered as Alternative Sources of feeding Akure

Two options were closely considered as alternative sources of electric power supply to Akure district distribution network. The consideration was based on factors such as proximity to load centres, present load capacity on the source under consideration, cost of implementation and time schedule of installation. The evaluation conducted revealed a tee-off from the Osogbo-Benin 330 kV transmission line as a preferred choice. This is because there is a surplus of 69,410,000 kWh on the line from which 152,552,000 kWh- an average monthly load demand for Akure metropolis-can be conveniently sourced. Moreover, a total sum of ₦1, 060,696,526 will be sufficient to execute the connection within a period of 6 months. This is cheap compared to the ₦ 2,741,573,575 required to connect Akure district distribution network to Osogbo in 20 months.

CHAPTER FIVE

CONCLUSION AND RECOMMENDATION

5.1 CONCLUSION

The study reveals the real and reactive power components of the Osogbo-Akure 132 kV transmission line at steady-state operating modes. At maximum operating mode, the values of real and reactive power components P_2 and Q_2 are 69.86 MW and 38.58 MVar respectively. The real and reactive power components P_2 and Q_2 at minimum operating mode are 48.68 MW and 26.26 MVar. The voltage output $V_2(kV)$ at maximum and minimum modes are 106.87 kV and 119.91 kV respectively. The efficiencies of the line at maximum and minimum operating modes were calculated to be 83.2% and 90.1% correspondingly while the percentage loading on the line at maximum mode was computed to be 63%.

This research also reveals the active and reactive components of technical power losses on the line ΔP and ΔQ as 14.14 MW and 22.99 MVar at maximum operating mode. ΔP and ΔQ at minimum operating mode was calculated as 5.32 MW and 8.65 correspondingly.

The most feasible and reliable alternative electric power source for Akure district distribution system was identified as a tee-off from the Osogbo-Benin 330 kV transmission line. Assessment reveals that Akure can be connected to this source for the cost of ₦1, 060,696,526 in 6 months. This will serve as an optional source of electricity in an event of a fault occurrence at the Osogbo-Akure 132 kV section of the National Control Centre, Osogbo.

A reliability assessment was computed on the operations of the Osogbo-Akure 132 kV transmission line between 2005 and 2009. A failure rate of 194 occurrences was observed in 2009. This translates to a failure rate of 45 minute per day. A Mean Time between Failures (MTBF) was calculated for 2009 as 43.74 hours per failure. Subsequently, a dynamic mathematical model, representing the trend line that the failure rates estimated for each of the years between 2005 and 2009 described, was developed. This equation is useful for predicting future failure rates on the radial line.

68

Thereafter, the Customer Oriented Interruption Indices-System Average Interruption Frequency Index (SAIFI), System Average Interruption Duration Index (SAIDI), Customer Average Interruption Frequency Index (CAIFI), Customer Average Interruption Duration Index (CAIDI), Average Service Availability Index (ASAI), Average Service Unavailability Index (ASUI)-were all computed to be 194, 274.87, 194, 1.42, 0.97, 0.03 respectively for 2009. These parameters are very useful for assessing the severity or significance of system failures in future reliability prediction analysis. They can also be used as a means of assessing the past performance of the system.

5.1.1. Contribution to Knowledge

This study provides:

a) voltage and efficiency profiles on the existing Osogbo-Akure 132 kV transmission line, which are relevant to transmission and primary distribution planning; and

b) the optimal conditions for adequate electric power supply to consumers in Akure district distribution network.

c) A dynamic model of the failure rates on the Osogbo-Akure 132 kV transmission line depicted by equation 4.1.

5.2 RECOMMENDATION

Power Holding Company of Nigeria (PHCN) should endeavour to implement a tee-off from the Osogbo-Akure 330 kV transmission line as an alternative source of electric power supply to Akure district distribution network. Ondo State resident, which is presently 3,500,000, is on the increase. Many in the diaspora are returning to the state and would require adequate electric power supply for their livelihood. Reliable electric power supply will raise the general living standard of the inhabitants of Akure and its environs by sustaining industrial activities and assisting domestic electrical appliances to work long with high technical-economic indices.

REFERENCES

Alex, A. K. (2008, March 26). *Category: Companies of Nigeria- National Electric Power Authority*. Retrieved 18:24:53 August 22, 2010, from Nigerian Wiki Web site: http://ww.nigerianwiki.com.

Allan, G. (2009, September 22). *Failure rate*. Retrieved 16:28:42, August 17, 2010, from Categories: Wikipedia, the free encyclopaedia: http//en.wikipedia.org

Billinton, R., & Allan, R. N. (2008). *Reliability Evaluation of Power Systems,* 2nd edition. New York: Springer Private Limited.

Cassaza, J., & Delea, F. (2003). *Understanding Electric Power Systems* (by the Institute of Electrical and Electronic Engineers). New Jersey: John Wiley and Sons, Inc.

Daniels, A. (2010, August 17). *Electrical Power in Nigeria - Overview*. Retrieved 20:29:37 August 17, 2010, from Electrical Power: Mbendi Information Services Web site: http://www.mbendi.com.

Dugan, R. C., McGranaghan, M. F., Surya, S., & Beaty, W. H. (2004). *Electrical Power Systems Quality*, 2nd Edition. www.digitallibrary.com: McGraw Hill.

El-Hawary, M. E. (2000). *Electrical Energy Systems*. (pp. 125-126). Boca Raton, Florida: CRC Press, LLC.

Grainger, J, Stevenson, W, (1994). *Power System Analysis*, New York: McGraw-Hill.

Grigsby, L. L. (2006). *Electric Power Engineering Handbook (Power Systems),* 2nd Edition. New York: Taylor and Francis Group.

Grigsby, L. L. (2007). *Electric Power Generation, Transmission and Distribution,* 2nd Edition. In Price W. W, *Electric Power Engineering Handbook* (pp. 10, chapter 13), CRC Press.

Grigsby, L. L. (2007). *Power System Stability and Control,* 2nd Edition. In Karady G. G, *Electric Power Engineering Handbook* (pp. 1, chapter 8), CRC Press.

Gupta, B. R. (2008). Power System Analysis and Design. (pp. 230-253). India: Rajendra Ravindra Printers, Ram Nagar, New Delhi.

Ifedi, V. (2010, July 30). *Power Holding Company of Nigeria*. Retrieved 20:45:22, August 17, 2010, from Article: Wikipedia-the free encyclopaedia Web site: http://en.wikipedia.org

Kothari, C. R. (2004). *Research Methodology*. 2nd Revised Edition. (pp. 134-135). India: Daharmesh Art Process.

Lakervi, E., & Holmes, E. J. (1989). *Electrical Distribution Network Design*. (pp. 70-82). England: Short Run Press Limited., Exeter.

Mehta, V., & Mehta, R. (2008). *Principles of Power Systems*, 4th Revised Edition. (pp. 127-130). info@schandgroup.com: S.Chand and Company Limited.

Meier, A. V. (2006). *Electric Power Systems*. (pp. 156-158, 166, 187, 229-233). Hoboken, New Jersey: John Wiley & Sons, Inc; A Wiley-Interscience publication.

Nagrath, I., & Kothari, D. (2001). *Power System Engineering* (p. 293). New Delhi: Tata McGraw-Hill Publishing Company Limited.

Pabla, A. (2005). *Electric Power Distribution*, Fifth Edition.(pp. 310-314). Delhi: Tata McGraw-Hill.

Reeves, E.A, Heathcote, Martin. J. (2003). *Newnes Electrical Pocket Book.* 23rd Edition. (pp.205). Chennai, India: Laserwords Private Limited.

Saccomanno, F. (2003). *Electric Power Systems - Analysis and Control*. (pp. 1-4). Hoboken, New Jersey: IEEE Press.

Sadaat, H. (2002). *Power System Analysis.* (pp. 208,240-242). India: Tata McGraw-Hill Publishing Company, Noida.

Siemens, A. (1998). Electrical Engineering Handbook, Third Edition. (pp. 314-316). New Delhi, India: Mohinder Singh Sejwal for Wiley Eastern Limited.

Shirley, B, Shirley, N. (2010, July 11). *Gauss-Seidel Power Flow Method.* Retrieved 20:36:11, September 28, 2010, from Discussion: Wikipedia, the free encyclopaedia: http//en.wikipedia.org

Short, T. (2004). *Electrical Power Distribution Handbook*. (pp. 44-45). Boca Raton, Florida: CRC Press LLC.

Theraja, B., & Theraja, A. (2008). A *Textbook of Electrical Technology*. (p. 510). India: S.Chand and company .

Weber, C. (2005). *Uncertainty in the Electric Power Industry*. New York: Springer Science + Business Media Inc.

APPENDIX A

APPENDIX A1: Half hourly Statistics of P (MW) and Q (MVar) for January, 2009

Minutes	MAX MW	MAX MVar	MEAN MW	MEAN MVar	STDVA, MW	STDVA, MVar	Phase angle of MVA (θ)	Power factor
0	56	31	39.39	18.52	11.85	6.58	29.06	0.87
00:30	56	31	38.87	18.29	11.27	6.56	30.19	0.86
01:00	62	31	39.77	18.29	11.46	6.53	29.68	0.87
01:30	56	30	38.77	18.65	11.13	6.64	30.83	0.86
02:00	55	30	38.03	17.87	10.29	6.59	32.63	0.84
02:30	55	30	37.68	17.58	10.29	6.62	32.75	0.84
03:00	55	30	37.48	17.32	9.87	6.26	32.39	0.84
03:30	55	30	37.65	17.00	10.24	6.20	31.20	0.86
04:00	55	29	36.67	16.83	10.61	6.28	30.60	0.86
04:30	55	29	37.36	17.46	10.86	6.74	31.82	0.85
05:00	55	30	38.13	17.58	10.59	6.63	32.04	0.85
05:30	55	30	38.97	18.00	9.99	6.04	31.18	0.86
06:00	56	29	40.50	19.23	13.07	7.42	29.58	0.87
06:30	65	29	44.42	20.65	11.12	6.33	29.64	0.87
07:00	63	36	42.57	19.90	12.26	7.81	32.51	0.84
07:30	68	35	41.13	19.39	13.28	7.74	30.22	0.86
08:00	61	37	40.26	19.90	11.31	7.62	33.95	0.83
08:30	70	36	40.74	20.32	12.92	7.54	30.29	0.86
09:00	61	38	41.48	21.23	11.39	8.18	35.68	0.81
09:30	60	36	39.94	21.29	10.83	7.66	35.26	0.82
10:00	60	37	38.83	20.93	10.34	7.45	35.77	0.81
10:30	60	38	37.31	20.28	11.40	7.46	33.19	0.84
11:00	54	37	36.87	20.58	10.26	7.42	35.89	0.81
11:30	55	34	35.57	19.20	10.31	6.79	33.37	0.84
12:00	53	32	35.10	18.68	10.91	7.29	33.74	0.83
12:30	54	31	35.39	18.74	11.10	7.10	32.63	0.84
13:00	56	34	35.96	19.18	13.38	8.68	32.96	0.84
13:30	57	39	34.07	18.13	13.92	9.50	34.33	0.83
14:00	56	32	33.13	17.27	13.73	8.20	30.83	0.86
14:30	56	32	33.57	17.82	13.49	7.64	29.52	0.87
15:00	56	32	31.90	16.69	13.81	7.74	29.25	0.87
15:30	53	29	32.80	17.00	12.72	6.70	27.75	0.88
16:00	52	29	32.03	15.97	13.11	7.59	30.08	0.87
16:30	50	29	33.93	16.83	11.71	7.42	32.38	0.84
17:00	47	27	34.23	16.77	11.50	6.98	31.26	0.85

17:30	48	26	37.47	18.67	8.67	5.59	32.82	0.84
18:00	52	28	40.29	20.19	8.18	5.74	35.03	0.82
18:30	60	35	42.32	21.84	8.99	6.26	34.85	0.82
19:00	60	37	42.81	23.11	13.00	7.23	29.07	0.87
19:30	63	39	44.37	22.47	11.03	8.72	38.33	0.78
20:00	84	38	47.00	22.00	12.65	7.43	30.44	0.86
20:30	60	34	46.47	22.33	10.08	6.81	34.03	0.83
21:00	60	38	46.32	22.06	9.99	7.17	35.69	0.81
21:30	62	34	46.32	23.03	9.65	7.03	36.05	0.81
22:00	61	37	46.57	23.70	9.89	7.34	36.58	0.80
22:30	63	36	46.48	23.87	9.11	6.94	37.30	0.80
23:00	59	38	43.84	22.06	9.37	7.51	38.70	0.78
23:30	58	32	43.48	21.74	8.79	6.27	35.52	0.81
24:00 :00	57	31	41.55	19.58	8.91	6.14	34.54	0.82

5.59

APPENDIX A2: Half hourly Statistics of P (MW) and Q (MVar) for February, 2009

Minutes	MAX MW	MAX MVar	MEAN MW	MEAN MVar	STDVA, MW	STDVA, MVar	Phase angle of (θ)	Power factor
0	68	31	41.45	19.00	12.30	7.11	30.03	0.87
00:30	57	31	39.61	18.26	9.19	6.51	35.30	0.82
01:00	55	29	38.87	18.30	8.70	5.67	33.09	0.84
01:30	55	26	38.30	17.91	9.83	5.07	27.28	0.89
02:00	58	29	39.35	18.09	8.05	4.99	31.78	0.85
02:30	58	29	38.96	17.78	8.14	5.13	32.25	0.85
03:00	52	27	38.22	17.83	7.48	5.36	35.60	0.81
03:30	52	27	38.09	17.39	7.39	4.87	33.37	0.84
04:00	52	28	38.26	17.78	6.93	4.74	34.36	0.83
04:30	52	28	38.61	18.26	6.49	4.88	36.94	0.80
05:00	53	28	39.57	18.48	7.71	5.37	34.84	0.82
05:30	54	30	39.22	18.04	8.03	5.73	35.51	0.81
06:00	56	32	40.35	18.48	8.26	6.32	37.40	0.79
06:30	55	30	39.09	17.00	8.63	5.49	32.43	0.84
07:00	57	30	39.09	17.09	9.38	6.31	33.93	0.83
07:30	61	35	39.52	17.35	8.85	6.52	36.37	0.81

08:00	61	42	40.30	20.65	9.78	8.17	39.89	0.77
08:30	61	36	38.09	18.22	11.04	7.88	35.51	0.81
09:00	61	38	40.78	20.65	10.03	7.61	37.19	0.80
09:30	60	36	41.65	21.30	9.66	7.50	37.82	0.79
10:00	60	37	40.57	21.30	9.57	7.88	39.48	0.77
10:30	60	38	40.17	20.83	10.56	7.51	35.40	0.82
11:00	55	37	37.87	20.70	11.33	8.78	37.75	0.79
11:30	56	34	37.95	20.00	11.31	7.85	34.79	0.82
12:00	56	36	38.87	20.26	11.27	7.76	34.56	0.82
12:30	54	31	36.82	19.27	8.88	6.18	34.83	0.82
13:00	56	34	36.70	20.09	10.57	7.41	35.03	0.82
13:30	57	39	38.35	21.35	8.47	7.61	41.94	0.74
14:00	56	32	40.86	21.14	8.22	6.60	38.76	0.78
14:30	56	32	41.29	21.14	7.56	6.30	39.83	0.77
15:00	56	32	38.32	19.47	10.82	7.14	33.43	0.83
15:30	55	33	38.05	19.52	8.78	6.42	36.18	0.81
16:00	55	32	36.95	18.43	8.41	6.33	36.98	0.80
16:30	62	25	39.10	19.05	8.53	4.41	27.33	0.89
17:00	54	31	37.71	19.10	7.68	5.10	33.57	0.83
17:30	54	31	38.24	19.05	7.72	5.66	36.25	0.81
18:00	55	31	41.36	20.64	6.37	5.21	39.24	0.77
18:30	60	33	42.41	19.82	9.00	6.68	36.58	0.80
19:00	61	37	39.95	19.64	13.38	8.40	32.13	0.85
19:30	63	38	39.73	18.23	12.45	9.13	36.26	0.81
20:00	63	38	41.00	17.52	13.92	8.47	31.33	0.85
20:30	60	34	40.27	17.68	12.65	8.17	32.85	0.84
21:00	60	38	39.71	17.71	12.24	8.59	35.06	0.82
21:30	62	34	41.18	19.14	11.83	8.88	36.89	0.80
22:00	62	36	41.79	19.83	13.81	10.21	36.47	0.80
22:30	63	36	41.70	19.61	12.16	8.99	36.47	0.80
23:00	59	34	40.27	18.73	11.63	8.90	37.43	0.79
23:30	58	32	42.13	19.30	9.62	7.59	38.30	0.78
24:00 :00	57	31	40.67	19.33	11.37	6.96	31.47	0.85

4.41

APPENDIX A3: Half hourly Statistics of P (MW) and Q (MVar) for March, 2009

Minutes	MAX MW	MAX MVar	MEAN MW	MEAN MVar	STDVA, MW	STDVA, MVar	Phase angle of (θ)	Power factor
0	58	32	41.23	20.00	14.13	9.24	33.18	0.84
00:30	58	48	41.10	21.13	13.27	9.95	36.85	0.80
01:00	56	31	38.87	18.87	13.52	8.72	32.80	0.84
01:30	56	31	38.58	18.58	13.47	8.52	32.32	0.85
02:00	55	30	38.35	18.32	13.37	8.42	32.21	0.85
02:30	55	30	38.81	18.45	13.19	8.42	32.54	0.84
03:00	55	30	38.00	18.48	13.34	7.98	30.89	0.86
03:30	55	30	38.71	18.90	13.44	7.93	30.52	0.86
04:00	55	30	38.59	18.48	13.34	8.35	32.04	0.85
04:30	55	30	38.77	18.74	13.38	8.32	31.88	0.85
05:00	57	30	38.26	17.81	13.20	8.14	31.66	0.85
05:30	55	31	39.39	18.48	13.34	8.73	33.20	0.84
06:00	59	47	42.50	21.03	12.11	9.79	38.95	0.78
06:30	57	33	41.37	19.60	11.78	8.42	35.56	0.81
07:00	57	31	36.67	16.37	13.82	8.68	32.13	0.85
07:30	59	33	35.39	15.94	13.99	8.72	31.94	0.85
08:00	61	34	35.45	16.58	13.08	8.42	32.77	0.84
08:30	70	32	39.93	18.83	15.37	8.28	28.31	0.88
09:00	57	33	39.58	20.16	13.33	8.35	32.06	0.85
09:30	54	31	38.26	20.06	11.87	8.04	34.11	0.83
10:00	53	33	37.00	19.87	12.29	8.44	34.46	0.82
10:30	54	32	37.16	20.00	11.72	8.04	34.44	0.82
11:00	54	33	39.42	21.23	10.47	7.86	36.89	0.80
11:30	55	32	39.42	21.32	9.84	7.24	36.33	0.81
12:00	54	33	38.65	20.81	10.60	7.96	36.90	0.80
12:30	51	31	39.66	21.24	10.09	7.47	36.50	0.80
13:00	54	33	40.41	22.21	9.36	6.67	35.49	0.81
13:30	54	33	40.60	22.63	8.23	5.49	33.70	0.83
14:00	55	33	41.41	22.97	9.15	6.66	36.06	0.81
14:30	55	33	40.03	21.77	10.75	7.85	36.14	0.81
15:00	55	33	40.23	22.29	11.05	7.83	35.32	0.82
15:30	55	34	39.21	21.86	11.73	8.51	35.96	0.81
16:00	56	35	38.90	20.73	12.47	8.90	35.52	0.81
16:30	56	35	40.39	22.39	11.93	8.38	35.08	0.82
17:00	56	35	38.00	20.59	13.28	9.09	34.39	0.83

17:30	54	33	41.00	21.55	10.52	6.93	33.37	0.84
18:00	54	33	42.67	22.73	8.77	6.09	34.78	0.82
18:30	56	35	45.73	24.20	7.35	6.17	40.03	0.77
19:00	61	35	44.21	22.79	12.13	7.96	33.30	0.84
19:30	60	34	44.77	21.97	11.16	8.14	36.11	0.81
20:00	84	38	46.47	21.77	16.88	8.45	26.59	0.89
20:30	60	35	45.19	21.97	14.37	8.96	31.96	0.85
21:00	60	36	42.32	20.55	14.15	9.50	33.86	0.83
21:30	60	36	43.00	21.42	14.32	9.10	32.43	0.84
22:00	61	37	43.83	22.14	14.51	10.12	34.90	0.82
22:30	60	35	43.90	22.65	14.91	9.90	33.58	0.83
23:00	59	35	43.48	21.84	14.70	10.55	35.66	0.81
23:30	58	33	42.42	21.03	14.34	9.46	33.41	0.83
24:00 :00	58	32	41.19	20.29	14.58	8.98	31.64	0.85

5.49

APPENDIX A4: Half hourly Statistics of P (MW) and Q (MVar) for April, 2009

Minutes	MAX MW	MAX MVar	MEAN MW	MEAN MVar	STDVA, MW	STDVA, MVar	Phase angle of (θ)	Power factor
0	58	32	44.10	22.03	9.37	6.40	34.33	0.83
00:30	58	32	43.07	21.64	9.76	6.36	33.08	0.84
01:00	58	31	42.10	20.70	9.03	6.14	34.21	0.83
01:30		32	42.13	20.50	11.94	5.76	25.76	0.90
02:00	58	31	41.20	20.33	8.52	5.70	33.79	0.83
02:30	58	31	39.40	19.33	11.26	6.74	30.90	0.86
03:00	57	31	40.00	19.37	8.86	5.89	33.63	0.83
03:30	57	31	40.23	19.43	8.58	5.62	33.25	0.84
04:00	57	31	40.21	19.54	8.94	5.93	33.57	0.83
04:30	57	31	39.75	19.89	10.28	6.10	30.68	0.86
05:00	57	31	40.39	20.21	11.06	6.48	30.36	0.86
05:30	55	31	42.69	21.00	9.13	6.69	36.25	0.81
06:00	56	33	43.93	21.79	8.99	6.49	35.85	0.81
06:30	59	34	43.57	21.82	9.26	6.64	35.66	0.81
07:00	62	34	43.62	21.62	9.88	7.07	35.60	0.81

07:30	57	32	43.31	21.34	9.05	6.43	35.41	0.82
08:00	66	30	43.31	21.52	9.51	6.13	32.80	0.84
08:30	56	32	43.17	22.69	8.00	5.63	35.13	0.82
09:00	55	33	41.90	22.38	7.66	6.09	38.48	0.78
09:30	53	32	41.86	22.86	8.05	5.98	36.62	0.80
10:00	53	33	40.66	22.24	9.26	6.75	36.10	0.81
10:30	52	32	40.21	21.93	7.45	5.52	36.56	0.80
11:00	50	31	39.82	21.68	7.26	5.24	35.84	0.81
11:30	52	31	39.68	21.86	7.86	5.65	35.70	0.81
12:00	54	32	39.55	21.59	8.86	6.18	34.88	0.82
12:30	54	32	41.03	22.55	8.00	5.73	35.60	0.81
13:00	55	34	40.21	22.48	11.15	6.97	32.03	0.85
13:30	53	33	41.48	23.03	8.09	5.91	36.16	0.81
14:00	55	34	41.30	22.41	9.44	6.68	35.30	0.82
14:30	55	34	42.61	23.79	9.71	7.22	36.62	0.80
15:00	55	34	43.50	24.14	7.02	5.65	38.82	0.78
15:30	55	34	43.44	24.33	6.68	5.01	36.88	0.80
16:00	58	34	44.00	24.56	7.45	5.47	36.31	0.81
16:30	56	35	42.64	23.43	8.22	5.51	33.81	0.83
17:00	56	35	41.07	21.57	9.87	6.66	33.99	0.83
17:30	57	35	42.43	22.89	9.61	6.58	34.40	0.83
18:00	58	35	42.38	23.10	11.68	7.31	32.02	0.85
18:30	63	41	47.21	26.61	8.51	7.48	41.29	0.75
19:00	66	41	50.19	27.26	8.52	7.26	40.43	0.76
19:30	66	40	48.04	24.96	8.88	7.17	38.92	0.78
20:00	66	40	45.97	22.59	10.44	7.77	36.65	0.80
20:30	62	36	45.57	22.04	11.32	7.33	32.90	0.84
21:00	65	38	47.57	23.50	9.36	7.38	38.24	0.79
21:30	61	37	48.69	24.76	8.28	6.94	39.99	0.77
22:00	62	38	50.46	26.54	7.16	6.25	41.11	0.75
22:30	62	38	47.79	25.04	8.50	7.09	39.84	0.77
23:00	60	38	46.07	24.14	10.54	8.12	37.62	0.79
23:30	59	33	45.31	22.90	8.96	6.77	37.07	0.80
24:00 :00	58	32	44.79	22.34	9.17	6.14	33.79	0.83
						0.75		

APPENDIX A5: Half hourly Statistics of P (MW) and Q (MVar) for May, 2009

Minutes	MAX MW	MAX MVar	MEAN MW	MEAN MVar	STDVA, MW	STDVA, MVar	Phase angle of (θ)	Power factor
0	59	32	45	22	10	6	33.25	0.84
00:30	57	32	43	21	11	7	30.35	0.86
01:00	55	32	43	21	9	6	32.86	0.84
01:30	57	32	43	21	10	6	30.73	0.86
02:00	57	31	41	20	12	7	30.47	0.86
02:30	57	31	41	20	13	7	29.34	0.87
03:00	56	31	38	19	15	8	27.37	0.89
03:30	56	31	39	19	13	7	28.60	0.88
04:00	56	35	40	20	12	7	29.48	0.87
04:30	57	35	41	21	12	7	29.86	0.87
05:00	57	36	42	21	12	7	30.98	0.86
05:30	58	36	43	22	13	8	31.29	0.85
06:00	61	36	46	22	11	8	36.39	0.81
06:30	61	34	46	23	10	7	36.64	0.80
07:00	62	36	46	23	11	8	34.66	0.82
07:30	60	35	42	21	14	8	30.55	0.86
08:00	62	37	42	22	11	8	34.98	0.82
08:30	64	38	42	22	11	8	35.83	0.81
09:00	69	36	41	21	11	7	33.29	0.84
09:30	59	33	41	21	10	7	34.95	0.82
10:00	57	34	41	21	10	8	37.19	0.80
10:30	55	32	41	22	10	7	35.82	0.81
11:00	54	31	41	23	9	7	36.56	0.80
11:30	55	32	42	24	9	6	36.17	0.81
12:00	56	31	41	23	9	6	35.71	0.81
12:30	50	30	40	22	8	6	36.40	0.80
13:00	54	33	40	22	7	6	37.00	0.80
13:30	54	33	41	22	7	5	35.11	0.82
14:00	54	33	39	22	10	6	31.49	0.85
14:30	54	35	41	22	11	7	32.55	0.84
15:00	55	33	43	24	9	6	33.84	0.83
15:30	56	33	43	24	9	6	35.66	0.81
16:00	55	33	43	24	9	7	35.93	0.81
16:30	57	35	44	24	10	7	35.69	0.81
17:00	59	38	44	24	10	7	35.56	0.81

17:30	68	38	45	24	13	7	29.46	0.87
18:00	68	38	42	22	16	10	31.28	0.85
18:30	64	39	43	23	15	10	33.55	0.83
19:00	65	40	47	24	16	11	34.80	0.82
19:30	63	40	46	24	14	10	36.09	0.81
20:00	60	35	43	21	14	9	32.83	0.84
20:30	60	35	46	23	12	8	33.55	0.83
21:00	60	34	45	22	13	9	33.25	0.84
21:30	60	34	44	21	12	8	34.03	0.83
22:00	61	34	45	23	13	8	32.01	0.85
22:30	63	34	46	23	12	8	33.73	0.83
23:00	61	50	45	23	11	9	38.40	0.78
23:30	60	33	42	22	11	8	33.83	0.83
24:00 :00	59	32	44	21	11	7	34.52	0.82

5

APPENDIX A6: Half hourly Statistics of P (MW) and Q (MVar) for June, 2009

Minutes	MAX MW	MAX MVar	MEAN MW	MEAN MVar	STDVA, MW	STDVA, MVar	Phase angle of (θ)	Power factor
0	56	30	38.00	17.80	12.57	6.75	28.24	0.88
00:30	56	30	37.71	17.79	12.72	7.16	29.36	0.87
01:00	53	28	37.03	17.23	11.85	6.94	30.33	0.86
01:30	53	28	36.33	17.13	11.47	6.91	31.06	0.86
02:00	52	28	35.80	16.70	11.82	6.93	30.38	0.86
02:30	52	28	35.27	16.40	11.44	6.70	30.34	0.86
03:00	52	27	35.07	16.03	11.39	6.49	29.67	0.87
03:30	52	27	35.27	16.13	11.47	6.76	30.49	0.86
04:00	50	26	35.23	16.20	11.45	6.64	30.09	0.87
04:30	50	26	36.21	16.71	12.22	7.02	29.87	0.87
05:00	55	29	37.10	17.24	12.57	7.48	30.74	0.86
05:30	57	31	38.20	17.70	13.63	7.56	29.01	0.87
06:00	60	35	39.03	19.17	16.52	9.43	29.73	0.87
06:30	65	35	40.55	18.90	14.32	8.02	29.26	0.87
07:00	63	35	39.90	18.77	13.47	7.73	29.84	0.87

07:30	68	31	39.14	18.31	12.41	6.59	27.95	0.88
08:00	55	32	37.07	18.47	11.95	6.94	30.16	0.86
08:30	55	32	38.21	19.57	11.09	6.96	32.09	0.85
09:00	54	31	38.28	20.14	11.06	7.06	32.55	0.84
09:30	54	31	37.47	19.97	10.95	6.97	32.46	0.84
10:00	53	32	36.62	19.69	11.22	7.35	33.22	0.84
10:30	53	32	36.79	20.04	10.94	6.97	32.52	0.84
11:00	52	31	35.63	19.70	10.09	6.61	33.22	0.84
11:30	52	31	34.76	19.31	10.86	7.15	33.36	0.84
12:00	51	32	35.11	19.22	10.83	7.31	34.02	0.83
12:30	51	30	33.41	17.67	10.72	6.89	32.74	0.84
13:00	51	30	33.64	17.89	11.43	7.26	32.41	0.84
13:30	51	30	34.54	18.57	12.00	7.95	33.53	0.83
14:00	49	30	35.04	18.93	12.07	7.71	32.56	0.84
14:30	50	30	36.35	19.85	10.40	6.21	30.82	0.86
15:00	52	30	35.82	18.93	10.82	7.02	33.00	0.84
15:30	50	30	35.03	19.24	12.48	7.55	31.19	0.86
16:00	53	33	35.56	19.26	13.59	8.17	31.02	0.86
16:30	54	34	39.00	20.24	10.08	7.09	35.12	0.82
17:00	56	33	39.59	19.96	10.25	7.06	34.56	0.82
17:30	55	35	41.88	22.58	7.89	6.18	38.11	0.79
18:00	56	34	42.03	21.72	8.27	6.01	36.00	0.81
18:30	57	33	43.17	22.48	8.97	6.23	34.77	0.82
19:00	59	34	44.33	22.44	9.41	6.14	33.11	0.84
19:30	56	39	43.79	22.14	9.00	6.19	34.54	0.82
20:00	63	38	47.34	24.28	12.12	8.08	33.69	0.83
20:30	65	37	45.38	22.86	12.98	7.25	29.18	0.87
21:00	58	33	44.07	21.83	12.30	6.37	27.38	0.89
21:30	58	33	44.24	21.79	10.11	7.05	34.87	0.82
22:00	58	33	42.17	20.87	12.70	7.56	30.76	0.86
22:30	58	33	42.13	20.80	11.87	7.08	30.81	0.86
23:00	55	30	39.55	19.28	13.32	7.00	27.72	0.89
23:30	55	30	39.33	19.20	12.90	6.92	28.21	0.88
24:00 :00	53	30	37.87	17.37	12.00	6.39	28.03	0.88

6.01

APPENDIX A7: Half hourly Statistics of P (MW) and Q (MVar) for July, 2009

Minu tes	MAX MW	MAX MVar	MEAN MW	MEAN MVar	STDVA , MW	STDVA, MVar	Phase angle of (θ)	Power factor
0	54	27	38.61	17.55	7.93	4.89	31.66	0.85
00:30	52	26	37.32	16.71	7.87	4.58	30.18	0.86
01:00	52	25	36.45	16.26	7.98	4.40	28.85	0.88
01:30	52	24	36.42	16.42	7.82	4.32	28.90	0.88
02:00	50	24	35.68	16.23	7.44	4.46	30.91	0.86
02:30	50	27	35.45	16.23	7.71	4.73	31.54	0.85
03:00	50	24	34.74	15.58	8.12	4.44	28.67	0.88
03:30	56	25	35.50	15.83	8.96	4.60	27.19	0.89
04:00	49	25	33.77	15.19	8.56	4.83	29.41	0.87
04:30	50	25	33.48	14.97	9.18	5.28	29.88	0.87
05:00	52	25	34.26	15.35	9.36	5.43	30.09	0.87
05:30	54	37	36.03	16.84	10.01	6.63	33.52	0.83
06:00	56	30	39.26	18.33	11.06	6.52	30.50	0.86
06:30	58	33	41.77	19.58	11.51	6.87	30.83	0.86
07:00	60	35	42.77	20.90	10.01	6.04	31.10	0.86
07:30	58	36	41.87	20.73	10.67	6.79	32.48	0.84
08:00	57	32	40.30	19.80	10.40	6.49	31.98	0.85
08:30	57	33	39.30	19.57	10.28	6.40	31.90	0.85
09:00	58	34	38.37	19.47	10.46	6.55	32.05	0.85
09:30	56	33	36.39	18.61	10.59	6.86	32.94	0.84
10:00	55	32	37.10	19.19	9.42	6.01	32.54	0.84
10:30	54	33	37.52	19.35	9.43	6.32	33.83	0.83
11:00	53	32	37.11	19.79	9.04	5.67	32.12	0.85
11:30	52	31	36.55	18.66	9.01	5.72	32.38	0.84
12:00	51	30	34.76	18.21	10.75	6.60	31.53	0.85
12:30	51	30	35.04	18.32	10.02	6.31	32.19	0.85
13:00	53	31	34.93	18.10	9.90	6.64	33.86	0.83
13:30	50	29	35.07	18.27	9.41	5.92	32.16	0.85
14:00	48	27	34.00	17.78	9.57	6.11	32.55	0.84
14:30	49	27	33.96	17.93	8.46	5.01	30.63	0.86
15:00	49	28	33.32	17.68	10.45	6.10	30.28	0.86
15:30	50	28	34.87	18.10	9.42	5.42	29.93	0.87
16:00	49	28	35.10	17.97	7.38	4.52	31.50	0.85
16:30	49	28	36.13	18.45	6.61	4.31	33.13	0.84
17:00	48	28	36.10	18.35	7.89	4.92	31.93	0.85

17:30	50	27	37.63	18.37	7.47	5.08	34.19	0.83
18:00	53	29	39.36	19.25	7.42	5.28	35.45	0.81
18:30	55	31	39.97	20.03	9.88	6.14	31.85	0.85
19:00	60	35	44.03	22.10	9.17	6.87	36.86	0.80
19:30	63	39	44.57	22.73	11.71	7.68	33.25	0.84
20:00	65	58	45.70	23.27	12.35	11.09	41.93	0.74
20:30	66	50	47.10	24.03	12.62	9.97	38.30	0.78
21:00	62	36	45.97	22.55	10.25	7.36	35.66	0.81
21:30	61	34	44.93	22.07	10.74	7.26	34.05	0.83
22:00	58	32	42.47	20.57	11.29	7.12	32.23	0.85
22:30	58	31	42.28	20.03	9.95	6.49	33.14	0.84
23:00	58	31	40.20	18.93	9.89	6.05	31.44	0.85
23:30	54	29	37.70	17.50	9.69	5.80	30.91	0.86
24:00 :00	60	27	38.87	17.40	9.62	5.29	28.80	0.88

4.31

APPENDIX A8: Half hourly Statistics of P (MW) and Q (MVar) for August, 2009

Minu tes	MAX MW	MAX MVar	MEAN MW	MEAN MVar	STDVA , MW	STDVA, MVar	Phase angle of (θ)	Power factor
0	95	32	43.74	19.90	15.84	6.87	23.45	0.92
00:30	57	31	40.53	19.33	11.74	6.78	30.01	0.87
01:00	70	29	40.94	18.71	12.77	6.61	27.37	0.89
01:30	56	29	38.65	18.45	10.86	6.29	30.08	0.87
02:00	55	30	36.71	17.39	11.26	6.85	31.31	0.85
02:30	56	30	36.71	17.29	11.26	6.48	29.92	0.87
03:00	55	29	36.47	17.10	10.77	6.33	30.43	0.86
03:30	55	29	37.33	17.53	10.49	6.04	29.92	0.87
04:00	53	27	36.74	16.97	9.78	5.86	30.93	0.86
04:30	55	28	36.17	17.34	11.85	6.59	29.09	0.87
05:00	57	29	37.26	17.42	12.63	6.99	28.98	0.87
05:30	57	37	39.84	19.00	11.62	7.89	34.19	0.83
06:00	59	42	44.32	22.36	11.29	8.54	37.12	0.80
06:30	59	33	43.13	20.42	10.24	7.01	34.40	0.83
07:00	60	35	41.68	20.03	11.19	7.67	34.43	0.82
07:30	59	35	41.23	19.97	11.52	7.53	33.18	0.84

08:00	62	37	39.33	19.27	10.96	7.30	33.64	0.83
08:30	64	37	39.30	19.10	11.12	7.64	34.49	0.82
09:00	79	33	41.07	20.60	11.23	6.18	28.83	0.88
09:30	56	33	38.45	19.94	9.78	6.50	33.58	0.83
10:00	54	32	36.63	19.63	11.29	7.27	32.76	0.84
10:30	53	32	37.48	20.19	10.38	6.39	31.61	0.85
11:00	55	32	37.14	20.14	10.45	6.70	32.64	0.84
11:30	56	33	38.13	20.83	9.76	6.31	32.91	0.84
12:00	144	33	41.32	20.71	21.07	6.01	15.93	0.96
12:30	50	30	38.86	20.83	7.79	5.27	34.08	0.83
13:00	52	31	39.23	21.45	8.01	5.48	34.38	0.83
13:30	52	32	39.58	21.68	7.74	5.38	34.79	0.82
14:00	53	32	40.55	22.03	8.07	5.51	34.31	0.83
14:30	54	35	39.62	21.66	10.08	7.08	35.08	0.82
15:00	55	32	39.34	21.55	11.12	7.11	32.59	0.84
15:30	56	34	40.23	22.16	10.71	6.83	32.53	0.84
16:00	56	35	41.23	22.67	9.82	6.83	34.81	0.82
16:30	57	35	41.65	22.42	9.51	6.72	35.25	0.82
17:00	59	38	37.93	19.38	12.12	8.24	34.20	0.83
17:30	68	38	41.14	21.14	11.64	6.91	30.68	0.86
18:00	68	38	41.30	20.90	10.92	6.66	31.35	0.85
18:30	64	39	43.42	22.23	9.33	6.88	36.41	0.80
19:00	65	40	45.33	23.07	11.44	8.64	37.05	0.80
19:30	62	36	47.07	23.34	9.35	7.34	38.12	0.79
20:00	220	58	51.61	23.42	32.89	9.64	16.34	0.96
20:30	200	50	52.27	23.80	29.68	8.52	16.02	0.96
21:00	60	34	46.41	22.31	9.19	6.73	36.20	0.81
21:30	60	37	47.50	23.67	9.23	6.98	37.09	0.80
22:00	62	38	48.97	25.52	10.16	7.16	35.16	0.82
22:30	63	38	48.77	24.53	9.70	7.33	37.08	0.80
23:00	61	34	46.97	23.48	9.68	7.00	35.88	0.81
23:30	60	33	45.60	22.20	8.66	6.51	36.95	0.80
24:00 :00	59	32	43.52	20.00	8.87	6.63	36.81	0.80

5.27

APPENDIX A9: Half hourly Statistics of P (MW) and Q (MVar) for September, 2009

Minutes	MAX MW	MAX MVar	MEAN MW	MEAN MVar	STDVA, MW	STDVA, MVar	Phase angle of (θ)	Power factor
0	55	30	40.33	19.13	10.90	6.89	32.29	0.85
00:30	55	48	39.36	19.50	10.79	8.79	39.16	0.78
01:00	54	29	37.60	17.63	10.62	6.97	33.26	0.84
01:30	54	29	37.50	17.63	10.85	6.77	31.97	0.85
02:00	54	29	35.77	16.67	11.87	7.27	31.48	0.85
02:30	53	29	35.20	16.33	11.60	7.04	31.25	0.85
03:00	53	29	34.40	16.20	11.81	6.26	27.90	0.88
03:30	55	29	35.27	16.43	12.42	6.33	27.02	0.89
04:00	53	29	34.83	16.30	12.03	6.68	29.03	0.87
04:30	51	28	35.00	16.43	12.06	6.65	28.89	0.88
05:00	51	28	35.21	16.24	11.80	6.42	28.56	0.88
05:30	53	30	35.63	16.53	12.88	7.14	29.00	0.87
06:00	56	33	37.77	17.60	13.09	8.04	31.54	0.85
06:30	59	34	37.13	16.83	13.10	8.18	32.00	0.85
07:00	56	31	37.52	17.28	11.95	7.63	32.56	0.84
07:30	53	29	35.93	16.13	9.81	6.39	33.09	0.84
08:00	53	28	35.43	16.50	10.17	7.08	34.84	0.82
08:30	54	30	37.18	18.39	9.78	6.98	35.52	0.81
09:00	51	30	36.33	18.37	9.71	6.55	33.99	0.83
09:30	51	30	36.43	18.63	8.84	6.50	36.33	0.81
10:00	54	32	36.87	19.13	9.36	7.08	37.11	0.80
10:30	55	32	36.53	18.40	9.28	7.24	37.95	0.79
11:00	51	31	36.43	19.13	8.53	6.48	37.22	0.80
11:30	53	31	37.69	19.52	8.76	6.96	38.49	0.78
12:00	51	36	36.62	19.48	9.29	7.28	38.09	0.79
12:30	50	30	36.15	18.67	9.57	6.64	34.78	0.82
13:00	54	33	36.77	19.87	10.67	7.77	36.06	0.81
13:30	54	33	37.33	20.13	10.13	7.18	35.34	0.82
14:00	55	33	37.97	20.07	10.77	7.76	35.78	0.81
14:30	55	33	38.83	20.34	10.01	7.06	35.19	0.82
15:00	55	33	37.77	19.97	12.12	8.05	33.59	0.83
15:30	55	33	37.76	20.00	11.36	7.70	34.12	0.83
16:00	55	33	39.69	21.00	10.34	7.25	35.02	0.82
16:30	55	33	39.90	21.20	8.79	6.34	35.80	0.81
17:00	51	29	38.97	20.21	9.23	6.19	33.86	0.83

17:30	66	30	41.63	20.96	9.62	5.87	31.42	0.85
18:00	55	32	41.26	20.89	9.55	6.72	35.12	0.82
18:30	58	35	41.60	20.77	12.26	7.87	32.70	0.84
19:00	62	37	44.89	22.46	11.89	8.35	35.06	0.82
19:30	56	37	42.79	21.17	11.94	7.92	33.53	0.83
20:00	60	36	40.97	19.10	13.15	8.83	33.87	0.83
20:30	60	35	42.38	19.69	12.58	8.01	32.51	0.84
21:00	58	33	41.07	18.77	12.34	8.52	34.63	0.82
21:30	58	33	42.41	20.14	11.33	8.24	36.03	0.81
22:00	58	33	42.80	19.93	11.74	7.62	33.01	0.84
22:30	60	35	42.03	20.20	11.65	7.80	33.80	0.83
23:00	58	50	40.77	20.50	11.81	9.19	37.88	0.79
23:30	57	33	40.07	19.20	11.64	7.23	31.84	0.85
24:00 :00	55	30	39.43	17.97	10.72	7.00	33.16	0.84

5.87

APPENDIX A10: Half hourly Statistics of P (MW) and Q (MVar) for October, 2009

Minutes	MAX MW	MAX MVar	MEAN MW	MEAN MVar	STDVA, MW	STDVA, MVar	Phase angle of (θ)	Power factor
0	57	30	39.52	18.45	12.28	7.36	30.94	0.86
00:30	57	32	38.41	17.86	12.00	7.49	31.99	0.85
01:00	52	32	37.33	17.27	11.00	7.20	33.19	0.84
01:30	57	32	37.33	17.30	11.36	7.12	32.06	0.85
02:00	57	31	36.57	17.13	11.29	6.92	31.50	0.85
02:30	57	31	36.30	16.90	10.96	6.78	31.76	0.85
03:00	56	31	36.10	16.67	10.71	6.48	31.18	0.86
03:30	56	31	36.07	16.57	10.63	6.45	31.25	0.85
04:00	56	30	35.87	16.40	10.75	6.37	30.65	0.86
04:30	56	30	36.50	16.67	11.05	6.84	31.76	0.85
05:00	57	31	36.76	16.83	11.45	7.24	32.30	0.85
05:30	57	31	39.13	17.87	11.74	7.43	32.33	0.84
06:00	61	36	41.97	19.66	10.34	7.25	35.04	0.82
06:30	61	32	42.24	19.79	10.17	7.24	35.43	0.81
07:00	62	33	40.37	19.03	13.36	7.96	30.79	0.86

07:30	61	35	38.41	18.28	14.45	8.90	31.62	0.85
08:00	61	37	39.30	19.43	12.87	8.83	34.44	0.82
08:30	63	38	39.60	20.17	12.80	8.59	33.87	0.83
09:00	61	38	38.79	20.21	12.77	8.79	34.55	0.82
09:30	60	36	39.43	20.30	12.43	8.09	33.04	0.84
10:00	60	37	38.37	20.60	11.26	8.08	35.67	0.81
10:30	60	38	38.23	20.70	10.80	7.58	35.07	0.82
11:00	54	37	37.14	20.83	9.89	7.75	38.08	0.79
11:30	55	34	36.48	19.62	10.37	7.34	35.28	0.82
12:00	56	31	37.03	19.48	11.11	7.14	32.72	0.84
12:30	54	34	36.25	19.75	10.49	7.60	35.93	0.81
13:00	56	34	38.07	20.59	8.71	6.09	34.95	0.82
13:30	57	39	38.76	21.45	8.69	6.61	37.25	0.80
14:00	56	32	39.62	21.12	8.61	6.09	35.30	0.82
14:30	56	32	39.00	20.64	8.68	6.13	35.22	0.82
15:00	56	32	38.11	19.93	9.12	6.65	36.09	0.81
15:30	53	31	38.36	20.07	9.32	5.91	32.37	0.84
16:00	54	31	38.14	19.71	10.14	6.77	33.73	0.83
16:30	52	31	39.54	20.54	7.10	5.20	36.24	0.81
17:00	54	32	38.82	20.04	8.37	5.92	35.28	0.82
17:30	50	29	40.85	21.07	6.83	5.03	36.36	0.81
18:00	53	30	40.76	20.97	10.31	7.19	34.91	0.82
18:30	60	32	42.90	22.00	9.98	7.00	35.04	0.82
19:00	60	37	44.32	22.71	10.97	8.20	36.77	0.80
19:30	63	38	45.69	23.00	10.65	8.09	37.22	0.80
20:00	63	38	45.45	21.86	11.78	8.53	35.92	0.81
20:30	61	37	46.38	23.07	11.21	8.03	35.61	0.81
21:00	60	38	47.03	23.30	9.95	7.07	35.38	0.82
21:30	62	34	45.87	22.84	12.54	8.29	33.48	0.83
22:00	61	37	44.81	22.32	12.29	8.63	35.07	0.82
22:30	63	36	44.20	22.17	11.36	8.67	37.34	0.80
23:00	59	38	41.27	20.27	13.26	9.35	35.20	0.82
23:30	58	32	41.37	21.07	11.65	8.40	35.81	0.81
24:00 :00	60	31	42.80	20.43	11.40	7.30	32.65	0.84

5.03

APPENDIX A11: Half hourly Statistics of P (MW) and Q (MVar) for November, 2009

Minu tes	MAX MW	MAX MVar	MEAN MW	MEAN MVar	STDVA , MW	STDVA, MVar	Phase angle of (θ)	Power factor
0	57	31	40.30	19.03	12.48	7.67	31.56	0.85
00:30	57	31	39.72	18.79	11.41	6.76	30.66	0.86
01:00	55	29	38.40	17.87	10.20	6.05	30.67	0.86
01:30	55	28	38.07	17.60	10.05	5.57	29.01	0.87
02:00	58	29	37.17	17.57	10.69	6.53	31.40	0.85
02:30	58	29	37.13	17.37	10.59	6.22	30.45	0.86
03:00	52	28	35.69	16.72	10.06	6.14	31.39	0.85
03:30	52	28	36.59	17.07	10.22	6.06	30.68	0.86
04:00	53	28	36.10	16.45	9.95	6.16	31.76	0.85
04:30	53	28	34.93	16.61	11.37	6.66	30.38	0.86
05:00	55	29	36.60	16.80	11.34	6.86	31.18	0.86
05:30	57	37	38.37	17.70	11.13	7.94	35.50	0.81
06:00	56	32	40.82	19.75	12.42	7.84	32.26	0.85
06:30	56	32	40.17	18.67	11.29	6.85	31.27	0.85
07:00	55	31	37.83	17.93	12.35	7.17	30.12	0.86
07:30	56	32	37.17	17.40	12.77	7.40	30.09	0.87
08:00	57	31	36.45	17.03	10.57	6.49	31.57	0.85
08:30	70	32	38.24	17.41	12.56	7.15	29.64	0.87
09:00	69	33	38.66	18.79	11.50	7.02	31.41	0.85
09:30	57	32	38.17	19.00	10.01	6.44	32.76	0.84
10:00	53	31	37.13	19.20	8.99	6.10	34.19	0.83
10:30	53	31	37.33	19.43	9.64	6.60	34.38	0.83
11:00	55	32	36.97	19.72	10.58	6.90	33.12	0.84
11:30	56	33	37.48	20.00	10.19	6.81	33.76	0.83
12:00	56	33	36.73	20.07	10.36	7.23	34.89	0.82
12:30	51	31	36.43	19.50	9.86	6.72	34.26	0.83
13:00	52	32	38.48	20.69	8.50	6.12	35.75	0.81
13:30	50	31	37.97	20.23	8.76	6.33	35.83	0.81
14:00	51	30	38.89	20.86	8.70	5.96	34.42	0.82
14:30	49	30	38.21	20.36	9.24	6.34	34.44	0.82
15:00	52	30	37.26	19.63	10.91	6.99	32.64	0.84
15:30	55	33	38.34	20.55	10.67	6.61	31.78	0.85
16:00	55	32	38.43	20.39	10.14	6.76	33.66	0.83
16:30	62	29	39.14	20.03	9.72	5.72	30.48	0.86
17:00	54	31	37.69	18.93	9.54	6.24	33.21	0.84

17:30	68	34	40.14	20.66	10.33	5.91	29.77	0.87
18:00	68	31	41.87	21.17	10.42	5.72	28.76	0.88
18:30	56	33	44.60	23.03	7.53	5.74	37.33	0.80
19:00	61	35	45.75	24.57	14.11	7.82	29.00	0.87
19:30	63	37	46.79	23.21	10.08	7.70	37.38	0.79
20:00	84	58	47.23	23.40	12.62	9.65	37.40	0.79
20:30	66	50	47.43	23.23	11.57	8.86	37.44	0.79
21:00	60	34	45.32	21.71	11.01	7.32	33.64	0.83
21:30	60	37	45.76	22.45	10.99	7.27	33.49	0.83
22:00	62	38	46.36	23.79	12.21	8.12	33.64	0.83
22:30	62	38	46.52	23.07	10.59	7.39	34.89	0.82
23:00	59	34	45.10	22.34	10.67	8.02	36.94	0.80
23:30	59	33	44.03	21.69	10.15	6.95	34.41	0.83
24:00 :00	57	31	41.63	19.40	11.07	6.97	32.21	0.85

5.57

APPENDIX A12: Half hourly Statistics of P (MW) and Q (MVar) for December, 2009

Minu tes	MAX MW	.MAX MVar	MEAN MW	MEAN MVar	STDVA , MW	STDVA, MVar	Phase angle of (θ)	Power factor
0	68	30	43.03	20.32	10.79	6.61	31.48	0.85
00:30	57	48	41.21	21.00	10.41	8.66	39.76	0.77
01:00	54	32	39.23	18.87	10.82	6.89	32.48	0.84
01:30	57	32	39.52	18.84	10.98	6.83	31.88	0.85
02:00	57	31	38.39	18.39	10.77	6.88	32.58	0.84
02:30	57	31	37.68	17.84	10.38	6.57	32.33	0.84
03:00	56	31	37.06	17.90	10.60	6.26	30.56	0.86
03:30	56	31	37.84	18.00	11.07	6.15	29.07	0.87
04:00	56	30	37.32	17.68	10.70	6.59	31.63	0.85
04:30	56	30	38.23	18.10	10.42	6.46	31.80	0.85
05:00	57	31	38.14	17.76	10.15	6.32	31.88	0.85
05:30	57	31	40.03	18.77	10.62	6.97	33.30	0.84
06:00	61	36	41.58	19.58	12.02	7.92	33.39	0.83
06:30	61	32	41.39	19.48	11.92	8.02	33.92	0.83
07:00	62	36	41.90	20.43	10.98	7.89	35.71	0.81
07:30	61	35	39.90	18.97	11.18	8.12	35.99	0.81

08:00	61	37	40.19	19.97	11.46	8.53	36.68	0.80
08:30	63	38	40.35	21.23	12.10	9.12	37.01	0.80
09:00	61	38	38.77	20.45	12.11	8.45	34.91	0.82
09:30	60	36	39.71	21.23	11.55	8.12	35.11	0.82
10:00	60	37	40.26	21.97	11.22	8.26	36.36	0.81
10:30	60	38	39.13	20.97	11.14	8.22	36.41	0.80
11:00	54	37	38.39	21.39	9.44	7.48	38.40	0.78
11:30	55	34	38.90	20.87	9.27	6.90	36.65	0.80
12:00	56	31	38.73	20.53	9.74	6.56	33.96	0.83
12:30	54	31	37.61	20.04	9.67	7.06	36.15	0.81
13:00	56	34	37.55	20.79	10.89	7.87	35.84	0.81
13:30	57	39	38.61	21.71	9.80	7.50	37.42	0.79
14:00	56	33	39.57	21.73	11.29	7.78	34.57	0.82
14:30	56	33	40.66	22.07	10.03	6.79	34.10	0.83
15:00	56	33	37.37	20.37	13.22	8.29	32.11	0.85
15:30	55	33	37.87	20.37	11.58	7.47	32.81	0.84
16:00	55	32	39.90	21.45	8.96	6.11	34.27	0.83
16:30	55	33	40.23	21.52	9.42	6.34	33.95	0.83
17:00	54	32	40.29	21.42	9.61	6.44	33.83	0.83
17:30	53	30	40.73	21.20	9.83	6.50	33.46	0.83
18:00	54	32	41.66	21.86	10.44	7.29	34.91	0.82
18:30	60	35	43.00	22.06	11.89	7.80	33.25	0.84
19:00	62	37	44.46	22.64	11.92	9.03	37.14	0.80
19:30	63	38	45.37	22.63	9.96	8.32	39.86	0.77
20:00	63	38	43.81	20.97	13.97	9.24	33.49	0.83
20:30	60	35	43.39	20.97	13.76	8.90	32.91	0.84
21:00	60	38	44.37	21.27	11.69	8.81	37.00	0.80
21:30	62	33	43.65	21.52	12.39	8.53	34.54	0.82
22:00	61	35	44.65	21.77	11.16	8.06	35.83	0.81
22:30	63	36	44.62	22.28	11.08	8.31	36.87	0.80
23:00	58	50	41.00	20.23	11.44	9.58	39.95	0.77
23:30	58	33	42.39	21.00	10.70	7.76	35.96	0.81
24:00 :00	57	31	41.45	19.94	10.63	7.07	33.62	0.83

6.11

APPENDIX B

APPENDIX B.1: Data of Electric Power Outages in 2009

Summary of Outages in 2009				
Date	Time Out	Time Restored	Frequency of outage	r (1 hour duration)
01/01/2009			0	0:00
02/01/2009	12:00	16:00	1	4:00
03/01/2009	2:00	3:00	1	1:00
04/01/2009			0	0:00
05/01/2009			0	0:00
06/01/2009			0	0:00
07/01/2009			0	0:00
07/01/2009	14:00	15:00	1	1:00
08/01/2009	9:00	13:00	1	4:00
09/01/2009	5:00	6:00	1	1:00
10/01/2009	16:00	17:00	1	1:00
11/01/2009			0	0:00
11/01/2009			0	0:00
11/01/2009			0	0:00
12/01/2009			0	0:00
13/01/2009	11:00	18:00	1	7:00
14/01/2009	8:00	9:00	1	1:00
15/01/2009			0	0:00
16/01/2009	19:00	21:00	1	2:00
16/01/2009			0	0:00
17/01/2009	2:00	3:00	1	1:00
17/01/2009			0	0:00
17/01/2009	19:00	20:00	1	1:00
18/01/2009			0	0:00
19/01/2009			0	0:00
19/01/2009			0	0:00
19/01/2009			0	0:00
19/01/2009			0	0:00
20/01/2009			0	0:00
21/01/2009			0	0:00
22/01/2009	1:00	2:00	1	1:00
23/01/2009	0:00	5:00	1	5:00
24/01/2009			0	0:00
25/01/2009	11:00	12:00	1	1:00
26/01/2009	12:00	13:00	1	1:00
27/01/2009			0	0:00
27/01/2009			0	0:00
28/01/2009			0	0:00

29/01/2009			0	0:00
30/01/2009			0	0:00
31/01/2009	12:00	16:00	1	4:00
31/01/2009	2:00	3:00	1	1:00
01/02/2009			0	0:00
02/02/2009			0	0:00
03/02/2009			0	0:00
04/02/2009			0	0:00
04/02/2009	14:00	15:00	1	1:00
05/02/2009	9:00	13:00	1	4:00
06/02/2009	5:00	6:00	1	1:00
06/02/2009	12:00	13:00	1	1:00
07/02/2009			0	0:00
08/02/2009			0	0:00
09/02/2009			0	0:00
10/02/2009			0	0:00
11/02/2009			0	0:00
12/02/2009			0	0:00
13/02/2009			0	0:00
13/02/2009			0	0:00
14/02/2009			0	0:00
15/02/2009			0	0:00
16/02/2009			0	0:00
17/02/2009	15:00	16:00	1	1:00
18/02/2009			0	0:00
19/02/2009			0	0:00
20/02/2009			0	0:00
21/02/2009	10:00	11:00	1	1:00
22/02/2009			0	0:00
23/02/2009			0	0:00
23/02/2009			0	0:00
24/02/2009	13:00	14:00	1	1:00
24/02/2009			0	0:00
24/02/2009	14:00	15:00	1	1:00
25/02/2009	16:00	18:00	1	2:00
26/02/2009			0	0:00
27/02/2009			0	0:00
28/02/2009	13:00	14:00	1	1:00
01/03/2009	15:00	16:00	1	1:00
02/03/2009			0	0:00
03/03/2009			0	0:00
04/03/2009	12:00	13:00	1	1:00
04/03/2009			0	0:00
05/03/2009	5:00	6:00	1	1:00
06/03/2009	19:00	20:00	1	1:00
07/03/2009	14:00	15:00	1	1:00

07/03/2009	16:00	17:00	1	1:00
08/03/2009	17:00	18:00	1	1:00
09/03/2009	11:00	12:00	1	1:00
10/03/2009	17:00	18:00	1	1:00
11/03/2009	23:00	23:59	1	0:59
12/03/2009	10:00	11:00	1	1:00
13/03/2009	14:00	15:00	1	1:00
13/03/2009	11:00	12:00	1	1:00
14/03/2009	15:00	16:00	1	1:00
15/03/2009	19:00	20:00	1	1:00
16/03/2009			0	0:00
16/03/2009			0	0:00
17/03/2009			0	0:00
17/03/2009	6:00	7:00	1	1:00
18/03/2009			0	0:00
19/03/2009	22:00	23:00	1	1:00
20/03/2009	23:00	23:59	1	0:59
21/03/2009			0	0:00
22/03/2009			0	0:00
23/03/2009			0	0:00
24/03/2009			0	0:00
25/03/2009	8:00	9:00	1	1:00
25/03/2009			0	0:00
26/03/2009			0	0:00
26/03/2009			0	0:00
27/03/2009	5:00	6:00	1	1:00
28/03/2009	14:00	15:00	1	1:00
29/03/2009			0	0:00
30/03/2009			0	0:00
31/03/2009			0	0:00
01/04/2009			0	0:00
02/04/2009			0	0:00
03/04/2009			0	0:00
04/04/2009			0	0:00
05/04/2009	14:00	15:00	1	1:00
06/04/2009	11:00	12:00	1	1:00
07/04/2009			0	0:00
07/04/2009	18:00	19:00	1	1:00
08/04/2009			0	0:00
09/04/2009			0	0:00
10/04/2009			0	0:00
11/04/2009			0	0:00
12/04/2009			0	0:00
13/04/2009			0	0:00
14/04/2009			0	0:00
15/04/2009			0	0:00

Date	Start	End	Count	Duration
16/04/2009			0	0:00
17/04/2009			0	0:00
18/04/2009	6:00	7:00	1	1:00
19/04/2009	12:00	14:00	1	2:00
20/04/2009	19:00	20:00	1	1:00
21/04/2009	14:00	15:00	1	1:00
21/04/2009	18:00	19:00	1	1:00
22/04/2009	6:00	7:00	1	1:00
22/04/2009	6:00	7:00	1	1:00
23/04/2009			0	0:00
23/04/2009			0	0:00
24/04/2009			0	0:00
25/04/2009			0	0:00
26/04/2009	11:00	12:00	1	1:00
26/04/2009	14:00	15:00	1	1:00
26/04/2009	21:00	22:00	1	1:00
27/04/2009	23:00	23:59	1	0:59
28/04/2009	3:00	4:00	1	1:00
29/04/2009	6:00	7:00	1	1:00
30/04/2009	8:00	10:00	1	2:00
30/04/2009	22:00	23:00	1	1:00
01/05/2009			0	0:00
02/05/2009			0	0:00
02/05/2009			0	0:00
03/05/2009			0	0:00
04/05/2009			0	0:00
05/05/2009			0	0:00
06/05/2009			0	0:00
06/05/2009			0	0:00
07/05/2009	22:00	23:00	1	1:00
08/05/2009	23:00	23:59	1	0:59
09/05/2009			0	0:00
09/05/2009			0	0:00
09/05/2009			0	0:00
10/05/2009			0	0:00
11/05/2009	8:00	9:00	1	1:00
12/05/2009			0	0:00
13/05/2009			0	0:00
14/05/2009			0	0:00
15/05/2009	5:00	6:00	1	1:00
16/05/2009	14:00	15:00	1	1:00
17/05/2009			0	0:00
18/05/2009			0	0:00
19/05/2009	13:00	14:00	1	1:00
19/05/2009	16:00	17:00	1	1:00
20/05/2009	11:00	13:00	1	2:00

Date	Start	End	Flag	Duration
21/05/2009	14:00	15:00	1	1:00
22/05/2009	9:00	13:00	1	4:00
23/05/2009	5:00	6:00	1	1:00
24/05/2009	12:00	13:00	1	1:00
25/05/2009			0	0:00
26/05/2009			0	0:00
27/05/2009			0	0:00
28/05/2009			0	0:00
28/05/2009			0	0:00
29/05/2009			0	0:00
29/05/2009			0	0:00
29/05/2009			0	0:00
30/05/2009			0	0:00
30/05/2009			0	0:00
31/05/2009			0	0:00
01/06/2009	15:00	16:00	1	1:00
02/06/2009			0	0:00
03/06/2009			0	0:00
04/06/2009			0	0:00
05/06/2009	10:00	11:00	1	1:00
06/06/2009			0	0:00
07/06/2009			0	0:00
08/06/2009			0	0:00
09/06/2009	13:00	14:00	1	1:00
10/06/2009			0	0:00
11/06/2009	14:00	15:00	1	1:00
11/06/2009	16:00	18:00	1	2:00
12/06/2009			0	0:00
13/06/2009			0	0:00
14/06/2009	13:00	14:00	1	1:00
14/06/2009	15:00	16:00	1	1:00
15/06/2009			0	0:00
16/06/2009			0	0:00
17/06/2009	12:00	13:00	1	1:00
18/06/2009			0	0:00
19/06/2009	5:00	6:00	1	1:00
19/06/2009	19:00	20:00	1	1:00
20/06/2009	14:00	15:00	1	1:00
20/06/2009	16:00	17:00	1	1:00
21/06/2009	17:00	18:00	1	1:00
22/06/2009	11:00	12:00	1	1:00
22/06/2009	17:00	18:00	1	1:00
22/06/2009	23:00	23:59	1	0:59
23/06/2009	10:00	11:00	1	1:00
23/06/2009	14:00	15:00	1	1:00
24/06/2009	11:00	12:00	1	1:00

24/06/2009	15:00	16:00	1	1:00
24/06/2009	19:00	20:00	1	1:00
25/06/2009	7:00	8:00	1	1:00
25/06/2009	13:00	14:00	1	1:00
25/06/2009	19:00	20:00	1	1:00
26/06/2009			0	0:00
27/06/2009	9:00	12:00	1	3:00
27/06/2009	15:00	16:00	1	1:00
28/06/2009	19:00	20:00	1	1:00
29/06/2009			0	0:00
30/06/2009	1:00	2:00	1	1:00
30/06/2009	6:00	7:00	1	1:00
01/07/2009			0	0:00
02/07/2009			0	0:00
03/07/2009	12:00	14:00	1	2:00
04/07/2009	20:00	21:00	1	1:00
05/07/2009			0	0:00
06/07/2009	6:00	7:00	1	1:00
07/07/2009	14:00	15:00	1	1:00
08/07/2009	11:00	12:00	1	1:00
09/07/2009			0	0:00
10/07/2009	18:00	19:00	1	1:00
11/07/2009			0	0:00
12/07/2009			0	0:00
13/07/2009			0	0:00
14/07/2009			0	0:00
15/07/2009			0	0:00
16/07/2009			0	0:00
17/07/2009			0	0:00
18/07/2009			0	0:00
19/07/2009			0	0:00
20/07/2009			0	0:00
21/07/2009	6:00	7:00	1	1:00
21/07/2009	12:00	14:00	1	2:00
21/07/2009	19:00	20:00	1	1:00
22/07/2009	14:00	15:00	1	1:00
22/07/2009	18:00	19:00	1	1:00
23/07/2009	6:00	7:00	1	1:00
24/07/2009	6:00	7:00	1	1:00
25/07/2009			0	0:00
26/07/2009			0	0:00
27/07/2009			0	0:00
28/07/2009			0	0:00
29/07/2009	11:00	12:00	1	1:00
29/07/2009	14:00	15:00	1	1:00
29/07/2009	21:00	22:00	1	1:00

29/07/2009	23:00	23:59	1	0:59
30/07/2009	3:00	4:00	1	1:00
30/07/2009	6:00	7:00	1	1:00
30/07/2009	8:00	10:00	1	2:00
30/07/2009	22:00	23:00	1	1:00
31/07/2009			0	0:00
01/08/2009			0	0:00
02/08/2009	6:00	7:00	1	1:00
03/08/2009			0	0:00
04/08/2009	22:00	23:00	1	1:00
05/08/2009	23:00	23:59	1	0:59
06/08/2009			0	0:00
07/08/2009			0	0:00
08/08/2009			0	0:00
09/08/2009			0	0:00
10/08/2009	8:00	9:00	1	1:00
11/08/2009			0	0:00
12/08/2009			0	0:00
13/08/2009			0	0:00
14/08/2009	5:00	6:00	1	1:00
14/08/2009	14:00	15:00	1	1:00
15/08/2009			0	0:00
16/08/2009			0	0:00
17/08/2009	13:00	14:00	1	1:00
17/08/2009	16:00	17:00	1	1:00
18/08/2009	11:00	13:00	1	2:00
18/08/2009	14:00	15:00	1	1:00
19/08/2009	9:00	13:00	1	4:00
20/08/2009	5:00	6:00	1	1:00
20/08/2009	12:00	13:00	1	1:00
21/08/2009	6:00	7:00	1	1:00
22/08/2009			0	0:00
23/08/2009			0	0:00
24/08/2009			0	0:00
25/08/2009			0	0:00
26/08/2009	13:00	17:00	1	4:00
26/08/2009	19:00	20:00	1	1:00
27/08/2009			0	0:00
28/08/2009			0	0:00
29/08/2009			0	0:00
30/08/2009	17:00	18:00	1	1:00
30/08/2009	20:00	21:00	1	1:00
31/08/2009	16:00	17:00	1	1:00
01/09/2009			0	0:00
02/09/2009	13:00	16:00	1	3:00
03/09/2009	18:00	19:00	1	1:00

04/09/2009	6:00	7:00	1	1:00
04/09/2009	9:00	10:00	1	1:00
05/09/2009			0	0:00
06/09/2009	4:00	5:00	1	1:00
07/09/2009			0	0:00
08/09/2009	20:00	21:00	1	1:00
09/09/2009			0	0:00
10/09/2009	5:00	9:00	1	4:00
11/09/2009			0	0:00
12/09/2009			0	0:00
13/09/2009	19:00	20:00	1	1:00
14/09/2009			0	0:00
15/09/2009	10:00	11:00	1	1:00
16/09/2009			0	0:00
17/09/2009			0	0:00
18/09/2009			0	0:00
19/09/2009			0	0:00
20/09/2009	14:00	15:00	1	1:00
21/09/2009			0	0:00
22/09/2009	17:00	18:00	1	1:00
23/09/2009			0	0:00
24/09/2009			0	0:00
25/09/2009	20:00	21:00	1	1:00
26/09/2009			0	0:00
27/09/2009			0	0:00
28/09/2009			0	0:00
29/09/2009			0	0:00
30/09/2009			0	0:00
01/10/2009			0	0:00
02/10/2009	13:00	14:00	1	1:00
03/10/2009			0	0:00
04/10/2009			0	0:00
05/10/2009	7:00	8:00	1	1:00
06/10/2009	12:00	14:00	1	2:00
07/10/2009			0	0:00
08/10/2009			0	0:00
09/10/2009			0	0:00
10/10/2009			0	0:00
11/10/2009	10:00	17:00	1	7:00
11/10/2009	18:00	19:00	1	1:00
12/10/2009	20:00	21:00	1	1:00
13/10/2009	20:00	23:59	1	3:59
14/10/2009	14:00	15:00	1	1:00
15/10/2009	16:00	17:00	1	1:00
15/10/2009	18:00	19:00	1	1:00
16/10/2009	5:00	6:00	1	1:00

16/10/2009	15:00	16:00	1	1:00
16/10/2009	18:00	19:00	1	1:00
16/10/2009	22:00	23:00	1	1:00
17/10/2009	21:00	22:00	1	1:00
18/10/2009			0	0:00
19/10/2009			0	0:00
20/10/2009			0	0:00
21/10/2009			0	0:00
22/10/2009			0	0:00
23/10/2009			0	0:00
24/10/2009			0	0:00
25/10/2009			0	0:00
26/10/2009			0	0:00
27/10/2009	18:00	19:00	1	1:00
28/10/2009			0	0:00
29/10/2009			0	0:00
30/10/2009	19:00	20:00	1	1:00
31/10/2009	10:00	19:00	1	9:00
01/11/2009			0	0:00
02/11/2009			0	0:00
03/11/2009	12:00	13:00	1	1:00
04/11/2009			0	0:00
05/11/2009	14:00	16:00	1	2:00
06/11/2009			0	0:00
07/11/2009			0	0:00
08/11/2009	18:00	19:00	1	1:00
09/11/2009			0	0:00
10/11/2009			0	0:00
11/11/2009			0	0:00
12/11/2009	21:00	22:00	1	1:00
13/11/2009			0	0:00
14/11/2009			0	0:00
15/11/2009			0	0:00
16/11/2009			0	0:00
17/11/2009			0	0:00
18/11/2009			0	0:00
19/11/2009			0	0:00
20/11/2009			0	0:00
21/11/2009			0	0:00
22/11/2009			0	0:00
23/11/2009			0	0:00
24/11/2009			0	0:00
25/11/2009			0	0:00
26/11/2009			0	0:00
27/11/2009			0	0:00
28/11/2009			0	0:00

Date				
29/11/2009			0	0:00
30/11/2009	13:00	14:00	1	1:00
30/11/2009	16:00	17:00	1	1:00
01/12/2009			0	0:00
02/12/2009			0	0:00
03/12/2009			0	0:00
04/12/2009			0	0:00
05/12/2009	11:00	18:00	1	7:00
06/12/2009	8:00	9:00	1	1:00
07/12/2009			0	0:00
08/12/2009	19:00	21:00	1	2:00
09/12/2009			0	0:00
10/12/2009	2:00	3:00	1	1:00
11/12/2009			0	0:00
12/12/2009	19:00	20:00	1	1:00
13/12/2009			0	0:00
14/12/2009			0	0:00
15/12/2009			0	0:00
16/12/2009			0	0:00
17/12/2009			0	0:00
18/12/2009			0	0:00
19/12/2009			0	0:00
20/12/2009	1:00	2:00	1	1:00
21/12/2009	0:00	5:00	1	5:00
22/12/2009			0	0:00
23/12/2009	11:00	12:00	1	1:00
24/12/2009	12:00	13:00	1	1:00
25/12/2009			0	0:00
26/12/2009			0	0:00
27/12/2009			0	0:00
28/12/2009			0	0:00
29/12/2009			0	0:00
30/12/2009	12:00	16:00	1	4:00
31/12/2009	2:00	3:00	1	1:00
			194	274.87

Source: Archive of Power Holding Company of Nigeria (PHCN) 132 kV Switching
Substation, Akure, Ondo State, Nigeria.

APPENDIX B.2: Data of Electric Power Outages in 2008

Summary of Outages in 2008				
Date	Time Out	Time Restored	Frequency of outage	r (1 hour duration)
01/01/2008			0	0:00
02/01/2008			0	0:00
03/01/2008			0	0:00
04/01/2008	9:00	10:00	1	1:00
05/01/2008			0	0:00
06/01/2008	10:00	12:00	1	2:00
06/01/2008	14:00	18:00	1	4:00
06/01/2008	20:00	21:00	1	1:00
07/01/2008			0	0:00
08/01/2008	13:00	14:00	1	1:00
09/01/2008	8:00	9:00	1	1:00
10/01/2008			0	0:00
11/01/2008	8:00	9:00	1	1:00
12/01/2008			0	0:00
13/01/2008	6:00	7:00	1	1:00
14/01/2008			0	0:00
15/01/2008	14:00	15:00	1	1:00
16/01/2008			0	0:00
17/01/2008	13:00	15:00	1	2:00
17/01/2008	18:00	19:00	1	1:00
18/01/2008	18:00	19:00	1	1:00
19/01/2008	22:00	23:00	1	1:00
20/01/2008			0	0:00
21/01/2008			0	0:00
22/01/2008	15:00	16:00	1	1:00
23/01/2008	14:00	17:00	1	3:00
23/01/2008	19:00	20:00	1	1:00
24/01/2008	15:00	16:00	1	1:00
25/01/2008			0	0:00
26/01/2008			0	0:00
27/01/2008	15:00	16:00	1	1:00
28/01/2008			0	0:00
29/01/2008	18:00	19:00	1	1:00
30/01/2008			0	0:00
31/01/2008			0	0:00
01/02/2008			0	0:00
02/02/2008			0	0:00
03/02/2008			0	0:00
04/02/2008			0	0:00
05/02/2008	6:00	7:00	1	1:00

			0	0:00
06/02/2008			0	0:00
07/02/2008	14:00	15:00	1	1:00
08/02/2008			0	0:00
09/02/2008	13:00	15:00	1	2:00
10/02/2008	18:00	19:00	1	1:00
11/02/2008	18:00	19:00	1	1:00
12/02/2008	22:00	23:00	1	1:00
13/02/2008			0	0:00
14/02/2008			0	0:00
15/02/2008			0	0:00
16/02/2008			0	0:00
17/02/2008			0	0:00
18/02/2008			0	0:00
19/02/2008			0	0:00
20/02/2008			0	0:00
21/02/2008			0	0:00
22/02/2008			0	0:00
23/02/2008			0	0:00
24/02/2008			0	0:00
25/02/2008			0	0:00
26/02/2008			0	0:00
27/02/2008			0	0:00
28/02/2008			0	0:00
29/02/2008			0	0:00
01/03/2008			0	0:00
02/03/2008			0	0:00
03/03/2008			0	0:00
04/03/2008			0	0:00
05/03/2008			0	0:00
06/03/2008			0	0:00
07/03/2008			0	0:00
08/03/2008			0	0:00
09/03/2008			0	0:00
10/03/2008			0	0:00
11/03/2008			0	0:00
12/03/2008			0	0:00
13/03/2008			0	0:00
14/03/2008			0	0:00
15/03/2008			0	0:00
16/03/2008			0	0:00
17/03/2008			0	0:00
18/03/2008	11:00	19:00	1	8:00
19/03/2008	15:00	16:00	1	1:00
20/03/2008	18:00	19:00	1	1:00
21/03/2008	21:00	22:00	1	1:00
22/03/2008			0	0:00

23/03/2008	15:00	18:00	1	3:00
24/03/2008			0	0:00
25/03/2008			0	0:00
26/03/2008	14:00	15:00	1	1:00
27/03/2008	21:00	22:00	1	1:00
28/03/2008	16:00	19:00	1	3:00
29/03/2008	16:00	17:00	1	1:00
30/03/2008	2:00	3:00	1	1:00
31/03/2008			0	0:00
01/04/2008			0	0:00
02/04/2008			0	0:00
03/04/2008			0	0:00
04/04/2008			0	0:00
05/04/2008			0	0:00
06/04/2008			0	0:00
07/04/2008			0	0:00
08/04/2008			0	0:00
09/04/2008			0	0:00
10/04/2008			0	0:00
11/04/2008			0	0:00
12/04/2008			0	0:00
13/04/2008			0	0:00
14/04/2008	10:00	11:00	1	1:00
15/04/2008	18:00	19:00	1	1:00
16/04/2008	22:00	23:00	1	1:00
17/04/2008			0	0:00
18/04/2008			0	0:00
19/04/2008			0	0:00
20/04/2008	9:00	10:00	1	1:00
21/04/2008	14:00	15:00	1	1:00
22/04/2008	20:00	22:00	1	2:00
23/04/2008	11:00	12:00	1	1:00
24/04/2008			0	0:00
25/04/2008			0	0:00
25/04/2008	16:00	18:00	1	2:00
26/04/2008	20:00	21:00	1	1:00
27/04/2008			0	0:00
27/04/2008	12:00	13:00	1	1:00
28/04/2008	14:00	15:00	1	1:00
28/04/2008	12:00	13:00	1	1:00
29/04/2008	14:00	15:00	1	1:00
30/04/2008	13:00	15:00	1	2:00
01/05/2008	16:00	17:00	1	1:00
02/05/2008	19:00	20:00	1	1:00
03/05/2008	11:00	19:00	1	8:00
03/05/2008	15:00	16:00	1	1:00

03/05/2008	18:00	19:00	1	1:00
04/05/2008	21:00	22:00	1	1:00
05/05/2008			0	0:00
06/05/2008	15:00	18:00	1	3:00
07/05/2008			0	0:00
08/05/2008			0	0:00
08/05/2008	14:00	15:00	1	1:00
09/05/2008	21:00	22:00	1	1:00
10/05/2008	16:00	19:00	1	3:00
11/05/2008	16:00	17:00	1	1:00
11/05/2008	2:00	3:00	1	1:00
11/05/2008	10:00	13:00	1	3:00
12/05/2008	18:00	20:00	1	2:00
13/05/2008	16:00	17:00	1	1:00
14/05/2008			0	0:00
15/05/2008			0	0:00
16/05/2008	1:00	3:00	1	2:00
17/05/2008	6:00	8:00	1	2:00
18/05/2008			0	0:00
19/05/2008	21:00	23:59	1	2:59
19/05/2008	0:00	1:00	1	1:00
20/05/2008	16:00	19:00	1	3:00
21/05/2008			0	0:00
22/05/2008			0	0:00
23/05/2008			0	0:00
24/05/2008			0	0:00
25/05/2008			0	0:00
26/05/2008			0	0:00
27/05/2008			0	0:00
28/05/2008			0	0:00
29/05/2008			0	0:00
30/05/2008	14:00	15:00	1	1:00
31/05/2008	1:00	4:00	1	3:00
01/06/2008	20:00	21:00	1	1:00
02/06/2008	11:00	23:59	1	12:59
03/06/2008	0:00	11:00	1	11:00
04/06/2008			0	0:00
05/06/2008	12:00	15:00	1	3:00
06/06/2008			0	0:00
07/06/2008			0	0:00
08/06/2008			0	0:00
09/06/2008			0	0:00
10/06/2008			0	0:00
11/06/2008	18:00	20:00	1	2:00
12/06/2008			0	0:00
13/06/2008			0	0:00

14/06/2008			0	0:00
15/06/2008			0	0:00
16/06/2008			0	0:00
17/06/2008			0	0:00
18/06/2008	17:00	18:00	1	1:00
19/06/2008			0	0:00
20/06/2008			0	0:00
21/06/2008			0	0:00
22/06/2008			0	0:00
23/06/2008			0	0:00
24/06/2008			0	0:00
25/06/2008			0	0:00
26/06/2008			0	0:00
27/06/2008			0	0:00
28/06/2008	9:00	11:00	1	2:00
29/06/2008			0	0:00
30/06/2008	14:00	15:00	1	1:00
30/06/2008	5:00	6:00	1	1:00
30/06/2008	8:00	9:00	1	1:00
30/06/2008	10:00	11:00	1	1:00
30/06/2008	12:00	15:00	1	3:00
01/07/2008	19:00	21:00	1	2:00
01/07/2008	4:00	5:00	1	1:00
02/07/2008	17:00	18:00	1	1:00
03/07/2008	14:00	15:00	1	1:00
04/07/2008	1:00	4:00	1	3:00
05/07/2008	20:00	21:00	1	1:00
06/07/2008	11:00	23:59	1	12:59
07/07/2008	0:00	11:00	1	11:00
08/07/2008			0	0:00
09/07/2008	12:00	15:00	1	3:00
10/07/2008			0	0:00
11/07/2008			0	0:00
12/07/2008			0	0:00
13/07/2008			0	0:00
14/07/2008			0	0:00
15/07/2008	18:00	20:00	1	2:00
16/07/2008			0	0:00
17/07/2008			0	0:00
18/07/2008			0	0:00
19/07/2008			0	0:00
20/07/2008			0	0:00
21/07/2008			0	0:00
22/07/2008	17:00	18:00	1	1:00
23/07/2008			0	0:00
24/07/2008	21:00	22:00	1	1:00

Date	Start	End	Count	Duration
25/07/2008			0	0:00
26/07/2008			0	0:00
27/07/2008	9:00	11:00	1	2:00
28/07/2008	18:00	19:00	1	1:00
28/07/2008	2:00	3:00	1	1:00
29/07/2008	19:00	20:00	1	1:00
30/07/2008			0	0:00
31/07/2008	10:00	11:00	1	1:00
01/08/2008			0	0:00
02/08/2008			0	0:00
03/08/2008			0	0:00
04/08/2008			0	0:00
05/08/2008			0	0:00
06/08/2008			0	0:00
07/08/2008	15:00	16:00	1	1:00
08/08/2008			0	0:00
09/08/2008			0	0:00
10/08/2008			0	0:00
11/08/2008	19:00	20:00	1	1:00
12/08/2008			0	0:00
13/08/2008			0	0:00
14/08/2008	16:00	18:00	1	2:00
15/08/2008	14:00	15:00	1	1:00
16/08/2008			0	0:00
17/08/2008			0	0:00
18/08/2008			0	0:00
19/08/2008			0	0:00
20/08/2008			0	0:00
21/08/2008	18:00	19:00	1	1:00
22/08/2008			0	0:00
22/08/2008	3:00	6:00	1	3:00
23/08/2008	18:00	19:00	1	1:00
24/08/2008			0	0:00
25/08/2008			0	0:00
26/08/2008			0	0:00
27/08/2008	12:00	14:00	1	2:00
28/08/2008	11:00	12:00	1	1:00
29/08/2008			0	0:00
30/08/2008			0	0:00
31/08/2008			0	0:00
01/09/2008			0	0:00
02/09/2008			0	0:00
03/09/2008	18:00	19:00	1	1:00
04/09/2008			0	0:00
05/09/2008	17:00	19:00	1	2:00
06/09/2008	16:00	17:00	1	1:00

07/09/2008	21:00	22:00	1	1:00
08/09/2008			0	0:00
09/09/2008			0	0:00
10/09/2008			0	0:00
11/09/2008			0	0:00
12/09/2008			0	0:00
13/09/2008			0	0:00
14/09/2008	18:00	19:00	1	1:00
15/09/2008			0	0:00
16/09/2008			0	0:00
17/09/2008	9:00	10:00	1	1:00
18/09/2008	17:00	18:00	1	1:00
19/09/2008	12:00	14:00	1	2:00
20/09/2008	16:00	17:00	1	1:00
21/09/2008			0	0:00
22/09/2008			0	0:00
23/09/2008			0	0:00
24/09/2008			0	0:00
25/09/2008			0	0:00
26/09/2008			0	0:00
27/09/2008			0	0:00
28/09/2008			0	0:00
29/09/2008			0	0:00
30/09/2008			0	0:00
01/10/2008			0	0:00
02/10/2008			0	0:00
03/10/2008			0	0:00
04/10/2008			0	0:00
05/10/2008			0	0:00
06/10/2008			0	0:00
07/10/2008			0	0:00
08/10/2008			0	0:00
09/10/2008			0	0:00
10/10/2008			0	0:00
11/10/2008			0	0:00
12/10/2008	15:00	16:00	1	1:00
13/10/2008			0	0:00
14/10/2008	10:00	19:00	1	9:00
14/10/2008	11:00	12:00	1	1:00
15/10/2008	14:00	15:00	1	1:00
16/10/2008			0	0:00
16/10/2008	11:00	13:00	1	2:00
16/10/2008	16:00	17:00	1	1:00
17/10/2008	18:00	19:00	1	1:00
18/10/2008			0	0:00
19/10/2008	17:00	19:00	1	2:00

19/10/2008	16:00	17:00	1	1:00
20/10/2008	21:00	22:00	1	1:00
21/10/2008			0	0:00
22/10/2008			0	0:00
23/10/2008			0	0:00
24/10/2008			0	0:00
25/10/2008			0	0:00
26/10/2008			0	0:00
27/10/2008	18:00	19:00	1	1:00
28/10/2008			0	0:00
29/10/2008			0	0:00
29/10/2008	9:00	10:00	1	1:00
30/10/2008	17:00	18:00	1	1:00
30/10/2008	12:00	14:00	1	2:00
31/10/2008	16:00	17:00	1	1:00
31/10/2008	7:00	10:00	1	3:00
01/11/2008	11:00	12:00	1	1:00
02/11/2008			0	0:00
02/11/2008	3:00	9:00	1	6:00
02/11/2008	12:00	13:00	1	1:00
03/11/2008	13:00	14:00	1	1:00
04/11/2008	5:00	6:00	1	1:00
05/11/2008			0	0:00
06/11/2008			0	0:00
07/11/2008	11:00	18:00	1	7:00
08/11/2008	19:00	20:00	1	1:00
09/11/2008			0	0:00
09/11/2008	16:00	17:00	1	1:00
10/11/2008	18:00	19:00	1	1:00
11/11/2008			0	0:00
12/11/2008	23:00	23:59	1	0:59
12/11/2008	15:00	16:00	1	1:00
12/11/2008	18:00	19:00	1	1:00
13/11/2008	21:00	22:00	1	1:00
14/11/2008	19:00	20:00	1	1:00
15/11/2008	12:00	14:00	1	2:00
16/11/2008			0	0:00
17/11/2008			0	0:00
18/11/2008	2:00	3:00	1	1:00
19/11/2008	22:00	23:00	1	1:00
20/11/2008			0	0:00
21/11/2008			0	0:00
22/11/2008			0	0:00
22/11/2008	18:00	19:00	1	1:00
23/11/2008	22:00	23:00	1	1:00
24/11/2008	18:00	19:00	1	1:00

24/11/2008	13:00	16:00	1	3:00
25/11/2008	17:00	22:00	1	5:00
26/11/2008			0	0:00
26/11/2008	13:00	14:00	1	1:00
27/11/2008	18:00	19:00	1	1:00
28/11/2008	16:00	17:00	1	1:00
29/11/2008	10:00	18:00	1	8:00
30/11/2008	18:00	21:00	1	3:00
01/12/2008	15:00	17:00	1	2:00
02/12/2008	12:00	14:00	1	2:00
03/12/2008	15:00	18:00	1	3:00
04/12/2008			0	0:00
05/12/2008	0:00	21:00	1	21:00
06/12/2008	13:00	14:00	1	1:00
07/12/2008			0	0:00
08/12/2008	1:00	2:00	1	1:00
09/12/2008			0	0:00
10/12/2008			0	0:00
10/12/2008	5:00	6:00	1	1:00
11/12/2008	12:00	13:00	1	1:00
11/12/2008	6:00	7:00	1	1:00
11/12/2008	8:00	9:00	1	1:00
11/12/2008	14:00	15:00	1	1:00
12/12/2008	19:00	20:00	1	1:00
13/12/2008	20:00	21:00	1	1:00
13/12/2008	11:00	12:00	1	1:00
14/12/2008	15:00	16:00	1	1:00
15/12/2008			0	0:00
16/12/2008			0	0:00
17/12/2008	14:00	15:00	1	1:00
17/12/2008	11:00	13:00	1	2:00
18/12/2008	14:00	15:00	1	1:00
19/12/2008			0	0:00
20/12/2008			0	0:00
21/12/2008			0	0:00
22/12/2008			0	0:00
22/12/2008	11:00	12:00	1	1:00
22/12/2008	13:00	14:00	1	1:00
23/12/2008	15:00	16:00	1	1:00
24/12/2008			0	0:00
25/12/2008	16:00	17:00	1	1:00
26/12/2008			0	0:00
27/12/2008	0:00	2:00	1	2:00
28/12/2008			0	0:00
29/12/2008			0	0:00
30/12/2008			0	0:00

31/12/2008			0	0:00
			184	356.93

Source: Archive of Power Holding Company of Nigeria (PHCN) 132 kV Switching
Substation, Akure, Ondo State, Nigeria.

APPENDIX B.3: Data of Electric Power Outages in 2007

Summary of Outages in 2007				
Date	Time Out	Time Restored	Frequency of outage	r (1 hour duration)
01/01/2007			0	0:00
02/01/2007			0	0:00
03/01/2007			0	0:00
04/01/2007	9:00	10:00	1	1:00
05/01/2007			0	0:00
06/01/2007			0	0:00
07/01/2007			0	0:00
08/01/2007			0	0:00
09/01/2007			0	0:00
10/01/2007			0	0:00
11/01/2007			0	0:00
12/01/2007			0	0:00
13/01/2007			0	0:00
14/01/2007			0	0:00
15/01/2007			0	0:00
16/01/2007			0	0:00
17/01/2007			0	0:00
18/01/2007			0	0:00
19/01/2007			0	0:00
20/01/2007			0	0:00
21/01/2007			0	0:00
22/01/2007			0	0:00
23/01/2007			0	0:00
24/01/2007			0	0:00
25/01/2007	20:00	21:00	1	1:00
26/01/2007	8:00	9:00	1	1:00
26/01/2007	18:00	19:00	1	1:00
27/01/2007	13:00	15:00	1	2:00
28/01/2007			0	0:00
29/01/2007	12:00	13:00	1	1:00
30/01/2007			0	0:00

31/01/2007			0	0:00
01/02/2007			0	0:00
02/02/2007	4:00	6:00	1	2:00
02/02/2007	21:00	23:00	1	2:00
03/02/2007			0	0:00
04/02/2007			0	0:00
05/02/2007			0	0:00
06/02/2007	10:00	12:00	1	2:00
07/02/2007	9:00	11:00	1	2:00
07/02/2007	14:00	15:00	1	1:00
08/02/2007	9:00	10:00	1	1:00
09/02/2007	17:00	19:00	1	2:00
10/02/2007			0	0:00
11/02/2007			0	0:00
12/02/2007	21:00	22:00	1	1:00
13/02/2007	14:00	15:00	1	1:00
13/02/2007	20:00	21:00	1	1:00
14/02/2007			0	0:00
15/02/2007	8:00	9:00	1	1:00
16/02/2007	15:00	17:00	1	2:00
16/02/2007	18:00	19:00	1	1:00
16/02/2007	22:00	23:00	1	1:00
17/02/2007	9:00	10:00	1	1:00
17/02/2007	15:00	16:00	1	1:00
17/02/2007	21:00	22:00	1	1:00
18/02/2007	6:00	7:00	1	1:00
18/02/2007	23:00	23:59	1	0:59
19/02/2007	21:00	22:00	1	1:00
20/02/2007	7:00	8:00	1	1:00
21/02/2007	6:00	7:00	1	1:00
22/02/2007	9:00	10:00	1	1:00
22/02/2007	14:00	15:00	1	1:00
23/02/2007			0	0:00
24/02/2007	17:00	19:00	1	2:00
24/02/2007	21:00	22:00	1	1:00
25/02/2007	10:00	11:00	1	1:00
25/02/2007	15:00	16:00	1	1:00
26/02/2007	6:00	7:00	1	1:00
26/02/2007	8:00	9:00	1	1:00
27/02/2007	17:00	18:00	1	1:00
28/02/2007			0	0:00
01/03/2007			0	0:00
02/03/2007	14:00	15:00	1	1:00
02/03/2007	18:00	19:00	1	1:00
02/03/2007	22:00	23:00	1	1:00
03/03/2007	8:00	9:00	1	1:00

03/03/2007	10:00	11:00	1	1:00
03/03/2007	22:00	23:00	1	1:00
04/03/2007	15:00	17:00	1	2:00
05/03/2007	14:00	15:00	1	1:00
05/03/2007	19:00	20:00	1	1:00
05/03/2007	21:00	22:00	1	1:00
06/03/2007	5:00	6:00	1	1:00
06/03/2007	14:00	15:00	1	1:00
06/03/2007	16:00	17:00	1	1:00
06/03/2007	19:00	20:00	1	1:00
06/03/2007	21:00	22:00	1	1:00
07/03/2007	7:00	8:00	1	1:00
07/03/2007	11:00	12:00	1	1:00
07/03/2007	15:00	16:00	1	1:00
08/03/2007	18:00	21:00	1	3:00
09/03/2007	11:00	14:00	1	3:00
10/03/2007			0	0:00
11/03/2007			0	0:00
12/03/2007	10:00	11:00	1	1:00
12/03/2007	16:00	18:00	1	2:00
13/03/2007	15:00	17:00	1	2:00
13/03/2007	21:00	22:00	1	1:00
14/03/2007	15:00	16:00	1	1:00
15/03/2007	16:00	17:00	1	1:00
15/03/2007	19:00	20:00	1	1:00
16/03/2007	11:00	13:00	1	2:00
16/03/2007	15:00	17:00	1	2:00
17/03/2007	6:00	7:00	1	1:00
18/03/2007	22:00	23:00	1	1:00
19/03/2007	15:00	16:00	1	1:00
20/03/2007	17:00	18:00	1	1:00
21/03/2007	9:00	10:00	1	1:00
21/03/2007	12:00	13:00	1	1:00
22/03/2007	17:00	18:00	1	1:00
22/03/2007	19:00	20:00	1	1:00
23/03/2007	19:00	20:00	1	1:00
24/03/2007	10:00	11:00	1	1:00
24/03/2007	18:00	19:00	1	1:00
25/03/2007	8:00	10:00	1	2:00
26/03/2007	17:00	18:00	1	1:00
27/03/2007	15:00	16:00	1	1:00
28/03/2007	4:00	5:00	1	1:00
28/03/2007	7:00	8:00	1	1:00
29/03/2007			0	0:00
30/03/2007	10:00	11:00	1	1:00
30/03/2007	18:00	19:00	1	1:00

31/03/2007	13:00	14:00	1	1:00
01/04/2007			0	0:00
02/04/2007			0	0:00
03/04/2007	6:00	7:00	1	1:00
04/04/2007	17:00	18:00	1	1:00
05/04/2007	11:00	12:00	1	1:00
05/04/2007	15:00	16:00	1	1:00
05/04/2007	18:00	19:00	1	1:00
06/04/2007	5:00	6:00	1	1:00
06/04/2007	15:00	16:00	1	1:00
06/04/2007	18:00	19:00	1	1:00
07/04/2007	17:00	18:00	1	1:00
08/04/2007	10:00	11:00	1	1:00
09/04/2007	13:00	14:00	1	1:00
09/04/2007	17:00	18:00	1	1:00
09/04/2007	19:00	20:00	1	1:00
09/04/2007	21:00	22:00	1	1:00
10/04/2007	1:00	2:00	1	1:00
10/04/2007	5:00	6:00	1	1:00
10/04/2007	19:00	20:00	1	1:00
11/04/2007			0	0:00
12/04/2007	15:00	16:00	1	1:00
13/04/2007			0	0:00
14/04/2007			0	0:00
15/04/2007			0	0:00
16/04/2007			0	0:00
17/04/2007	16:00	17:00	1	1:00
18/04/2007			0	0:00
19/04/2007			0	0:00
20/04/2007	0:00	1:00	1	1:00
21/04/2007			0	0:00
22/04/2007			0	0:00
23/04/2007			0	0:00
24/04/2007	12:00	13:00	1	1:00
24/04/2007	18:00	19:00	1	1:00
25/04/2007			0	0:00
26/04/2007			0	0:00
27/04/2007			0	0:00
28/04/2007			0	0:00
29/04/2007			0	0:00
30/04/2007			0	0:00
01/05/2007	18:00	19:00	1	1:00
02/05/2007	23:00	23:59	1	0:59
03/05/2007	0:00	1:00	1	1:00
03/05/2007	15:00	16:00	1	1:00
04/05/2007			0	0:00

			0	0:00
05/05/2007			0	0:00
06/05/2007	13:00	14:00	1	1:00
07/05/2007	11:00	13:00	1	2:00
07/05/2007	17:00	18:00	1	1:00
08/05/2007	9:00	17:00	1	8:00
09/05/2007	10:00	12:00	1	2:00
09/05/2007	17:00	18:00	1	1:00
10/05/2007	10:00	11:00	1	1:00
10/05/2007	17:00	18:00	1	1:00
11/05/2007			0	0:00
12/05/2007			0	0:00
13/05/2007			0	0:00
14/05/2007			0	0:00
15/05/2007			0	0:00
16/05/2007	16:00	17:00	1	1:00
17/05/2007			0	0:00
18/05/2007			0	0:00
19/05/2007			0	0:00
20/05/2007	1:00	2:00	1	1:00
20/05/2007	4:00	14:00	1	10:00
21/05/2007			0	0:00
22/05/2007	12:00	13:00	1	1:00
22/05/2007	15:00	16:00	1	1:00
23/05/2007	6:00	7:00	1	1:00
24/05/2007	4:00	5:00	1	1:00
24/05/2007	15:00	16:00	1	1:00
24/05/2007	23:00	23:59	1	0:59
25/05/2007			0	0:00
26/05/2007	23:00	23:59	1	0:59
27/05/2007	0:00	23:59	1	23:59
28/05/2007	0:00	15:00	1	15:00
28/05/2007	18:00	19:00	1	1:00
29/05/2007	21:00	22:00	1	1:00
30/05/2007			0	0:00
31/05/2007	5:00	6:00	1	1:00
01/06/2007	17:00	18:00	1	1:00
02/06/2007	12:00	13:00	1	1:00
03/06/2007			0	0:00
04/06/2007	12:00	14:00	1	2:00
05/06/2007			0	0:00
06/06/2007	9:00	10:00	1	1:00
07/06/2007			0	0:00
08/06/2007			0	0:00
09/06/2007			0	0:00
10/06/2007	10:00	11:00	1	1:00
11/06/2007			0	0:00

12/06/2007			0	0:00
13/06/2007			0	0:00
14/06/2007			0	0:00
15/06/2007	10:00	11:00	1	1:00
16/06/2007	9:00	11:00	1	2:00
16/06/2007	17:00	19:00	1	2:00
17/06/2007			0	0:00
18/06/2007			0	0:00
19/06/2007			0	0:00
20/06/2007	23:00	23:59	1	0:59
21/06/2007	0:00	15:00	1	15:00
22/06/2007			0	0:00
23/06/2007			0	0:00
24/06/2007			0	0:00
25/06/2007	16:00	17:00	1	1:00
26/06/2007			0	0:00
27/06/2007	20:00	21:00	1	1:00
28/06/2007	16:00	17:00	1	1:00
28/06/2007	20:00	21:00	1	1:00
29/06/2007	18:00	19:00	1	1:00
30/06/2007	13:00	17:00	1	4:00
01/07/2007			0	0:00
02/07/2007	12:00	17:00	1	5:00
02/07/2007	19:00	20:00	1	1:00
03/07/2007	15:00	16:00	1	1:00
04/07/2007	8:00	9:00	1	1:00
04/07/2007	10:00	11:00	1	1:00
04/07/2007	21:00	23:00	1	2:00
05/07/2007	16:00	17:00	1	1:00
06/07/2007	11:00	18:00	1	7:00
07/07/2007	2:00	4:00	1	2:00
07/07/2007	15:00	16:00	1	1:00
08/07/2007			0	0:00
09/07/2007	8:00	9:00	1	1:00
09/07/2007	14:00	15:00	1	1:00
10/07/2007			0	0:00
11/07/2007			0	0:00
12/07/2007	9:00	10:00	1	1:00
12/07/2007	15:00	20:00	1	5:00
13/07/2007	0:00	1:00	1	1:00
13/07/2007	21:00	23:59	1	2:59
14/07/2007	0:00	23:59	1	23:59
15/07/2007	0:00	23:59	1	23:59
16/07/2007	0:00	18:00	1	18:00
17/07/2007	14:00	16:00	1	2:00
18/07/2007	11:00	13:00	1	2:00

18/07/2007	17:00	18:00	1	1:00
19/07/2007			0	0:00
20/07/2007			0	0:00
21/07/2007			0	0:00
22/07/2007			0	0:00
23/07/2007			0	0:00
24/07/2007			0	0:00
25/07/2007			0	0:00
26/07/2007	10:00	12:00	1	2:00
27/07/2007			0	0:00
28/07/2007			0	0:00
29/07/2007			0	0:00
30/07/2007			0	0:00
31/07/2007			0	0:00
01/08/2007			0	0:00
02/08/2007			0	0:00
03/08/2007			0	0:00
04/08/2007	10:00	14:00	1	4:00
04/08/2007	17:00	18:00	1	1:00
05/08/2007	17:00	18:00	1	1:00
06/08/2007	12:00	14:00	1	2:00
07/08/2007	7:00	8:00	1	1:00
08/08/2007			0	0:00
09/08/2007			0	0:00
10/08/2007			0	0:00
11/08/2007	10:00	11:00	1	1:00
12/08/2007			0	0:00
13/08/2007			0	0:00
14/08/2007			0	0:00
15/08/2007			0	0:00
16/08/2007			0	0:00
17/08/2007			0	0:00
18/08/2007	9:00	11:00	1	2:00
18/08/2007	18:00	20:00	1	2:00
19/08/2007			0	0:00
20/08/2007			0	0:00
21/08/2007			0	0:00
22/08/2007			0	0:00
23/08/2007	0:00	1:00	1	1:00
24/08/2007	10:00	13:00	1	3:00
24/08/2007	14:00	16:00	1	2:00
24/08/2007	17:00	21:00	1	4:00
25/08/2007	19:00	23:59	1	4:59
26/08/2007	0:00	4:00	1	4:00
27/08/2007			0	0:00
28/08/2007	6:00	13:00	1	7:00

29/08/2007			0	0:00
30/08/2007			0	0:00
31/08/2007	13:00	14:00	1	1:00
01/09/2007			0	0:00
02/09/2007			0	0:00
03/09/2007			0	0:00
04/09/2007	14:00	16:00	1	2:00
05/09/2007	22:00	23:00	1	1:00
06/09/2007	16:00	17:00	1	1:00
06/09/2007	19:00	20:00	1	1:00
07/09/2007	20:00	23:00	1	3:00
08/09/2007			0	0:00
09/09/2007			0	0:00
10/09/2007	7:00	8:00	1	1:00
11/09/2007			0	0:00
12/09/2007			0	0:00
13/09/2007	6:00	7:00	1	1:00
14/09/2007	19:00	20:00	1	1:00
15/09/2007	1:00	2:00	1	1:00
15/09/2007	10:00	11:00	1	1:00
15/09/2007	17:00	19:00	1	2:00
16/09/2007			0	0:00
17/09/2007	14:00	15:00	1	1:00
17/09/2007	19:00	20:00	1	1:00
18/09/2007	7:00	8:00	1	1:00
19/09/2007	6:00	7:00	1	1:00
20/09/2007			0	0:00
21/09/2007			0	0:00
22/09/2007	23:00	23:59	1	0:59
23/09/2007			0	0:00
24/09/2007	16:00	19:00	1	3:00
25/09/2007	12:00	13:00	1	1:00
25/09/2007	17:00	18:00	1	1:00
26/09/2007	13:00	16:00	1	3:00
27/09/2007			0	0:00
28/09/2007			0	0:00
29/09/2007	11:00	12:00	1	1:00
29/09/2007	16:00	17:00	1	1:00
30/09/2007	16:00	19:00	1	3:00
01/10/2007	20:00	21:00	1	1:00
02/10/2007			0	0:00
03/10/2007			0	0:00
04/10/2007	18:00	19:00	1	1:00
05/10/2007			0	0:00
06/10/2007	9:00	12:00	1	3:00
06/10/2007	18:00	20:00	1	2:00

07/10/2007	14:00	16:00	1	2:00
07/10/2007	17:00	19:00	1	2:00
08/10/2007	17:00	18:00	1	1:00
09/10/2007	22:00	23:59	1	1:59
10/10/2007	0:00	9:00	1	9:00
10/10/2007	11:00	12:00	1	1:00
11/10/2007			0	0:00
12/10/2007			0	0:00
13/10/2007	7:00	8:00	1	1:00
14/10/2007			0	0:00
15/10/2007			0	0:00
16/10/2007	9:00	10:00	1	1:00
16/10/2007	15:00	16:00	1	1:00
17/10/2007			0	0:00
18/10/2007			0	0:00
19/10/2007			0	0:00
20/10/2007	22:00	23:59	1	1:59
21/10/2007	0:00	1:00	1	1:00
21/10/2007	7:00	8:00	1	1:00
21/10/2007	17:00	18:00	1	1:00
22/10/2007			0	0:00
23/10/2007	10:00	11:00	1	1:00
23/10/2007	17:00	18:00	1	1:00
24/10/2007	18:00	19:00	1	1:00
25/10/2007			0	0:00
26/10/2007	11:00	12:00	1	1:00
26/10/2007	14:00	15:00	1	1:00
27/10/2007			0	0:00
28/10/2007			0	0:00
29/10/2007			0	0:00
30/10/2007			0	0:00
31/10/2007	11:00	13:00	1	2:00
31/10/2007	18:00	19:00	1	1:00
01/11/2007	11:00	12:00	1	1:00
02/11/2007			0	0:00
03/11/2007	15:00	19:00	1	4:00
04/11/2007			0	0:00
05/11/2007	13:00	14:00	1	1:00
06/11/2007			0	0:00
07/11/2007	15:00	17:00	1	2:00
08/11/2007			0	0:00
09/11/2007	13:00	14:00	1	1:00
09/11/2007	17:00	18:00	1	1:00
10/11/2007			0	0:00
11/11/2007			0	0:00
12/11/2007			0	0:00

13/11/2007			0	0:00
14/11/2007			0	0:00
15/11/2007			0	0:00
16/11/2007			0	0:00
17/11/2007			0	0:00
18/11/2007			0	0:00
19/11/2007			0	0:00
20/11/2007			0	0:00
21/11/2007			0	0:00
22/11/2007			0	0:00
23/11/2007			0	0:00
24/11/2007			0	0:00
25/11/2007			0	0:00
26/11/2007			0	0:00
27/11/2007			0	0:00
28/11/2007			0	0:00
29/11/2007			0	0:00
30/11/2007			0	0:00
01/12/2007			0	0:00
02/12/2007	10:00	20:00	1	10:00
03/12/2007			0	0:00
04/12/2007			0	0:00
05/12/2007			0	0:00
06/12/2007			0	0:00
07/12/2007			0	0:00
08/12/2007			0	0:00
09/12/2007			0	0:00
10/12/2007	10:00	11:00	1	1:00
11/12/2007			0	0:00
12/12/2007			0	0:00
13/12/2007	14:00	17:00	1	3:00
14/12/2007			0	0:00
15/12/2007			0	0:00
16/12/2007			0	0:00
17/12/2007			0	0:00
18/12/2007			0	0:00
19/12/2007			0	0:00
20/12/2007			0	0:00
21/12/2007			0	0:00
22/12/2007			0	0:00
23/12/2007			0	0:00
24/12/2007			0	0:00
25/12/2007			0	0:00
26/12/2007			0	0:00
27/12/2007			0	0:00
28/12/2007			0	0:00

29/12/2007	1:00	2:00	1	1:00
30/12/2007			0	0:00
31/12/2007			0	0:00
			246	536.78

Source: Archive of Power Holding Company of Nigeria (PHCN) 132 kV Switching
Substation, Akure, Ondo State, Nigeria.

APPENDIX B.4: Data of Electric Power Outages in 2006

Summary of Outages in 2006				
Date	Time Out	Time Restored	Frequency of outage	r (1 hour duration)
01/01/2006			0	0:00
02/01/2006	11:00	12:00	1	1:00
03/01/2006			0	0:00
04/01/2006			0	0:00
05/01/2006			0	0:00
06/01/2006			0	0:00
07/01/2006			0	0:00
08/01/2006			0	0:00
09/01/2006			0	0:00
10/01/2006			0	0:00
11/01/2006			0	0:00
12/01/2006			0	0:00
13/01/2006			0	0:00
14/01/2006			0	0:00
15/01/2006			0	0:00
16/01/2006	4:00	5:00	1	1:00
16/01/2006	13:00	14:00	1	1:00
17/01/2006			0	0:00
18/01/2006			0	0:00
19/01/2006			0	0:00
20/01/2006			0	0:00
21/01/2006			0	0:00
22/01/2006			0	0:00
23/01/2006			0	0:00
24/01/2006			0	0:00
25/01/2006			0	0:00
26/01/2006			0	0:00
27/01/2006	11:00	13:00	1	2:00
28/01/2006			0	0:00

29/01/2006			0	0:00
30/01/2006			0	0:00
31/01/2006			0	0:00
01/02/2006			0	0:00
02/02/2006			0	0:00
03/02/2006			0	0:00
04/02/2006			0	0:00
05/02/2006			0	0:00
06/02/2006			0	0:00
07/02/2006			0	0:00
08/02/2006	17:00	18:00	1	1:00
09/02/2006			0	0:00
10/02/2006			0	0:00
11/02/2006	11:00	12:00	1	1:00
12/02/2006			0	0:00
13/02/2006			0	0:00
14/02/2006			0	0:00
15/02/2006	1:00	2:00	1	1:00
16/02/2006			0	0:00
17/02/2006			0	0:00
18/02/2006			0	0:00
19/02/2006			0	0:00
20/02/2006	8:00	9:00	1	1:00
21/02/2006			0	0:00
22/02/2006			0	0:00
23/02/2006			0	0:00
24/02/2006			0	0:00
25/02/2006			0	0:00
26/02/2006	6:00	8:00	1	2:00
26/02/2006	8:00	9:00	1	1:00
27/02/2006	17:00	18:00	1	1:00
28/02/2006			0	0:00
01/03/2006	14:00	16:00	1	2:00
02/03/2006			0	0:00
03/03/2006			0	0:00
04/03/2006			0	0:00
05/03/2006			0	0:00
06/03/2006			0	0:00
07/03/2006			0	0:00
08/03/2006			0	0:00
09/03/2006			0	0:00
10/03/2006			0	0:00
11/03/2006			0	0:00
12/03/2006			0	0:00
13/03/2006			0	0:00
14/03/2006			0	0:00

15/03/2006			0	0:00
16/03/2006			0	0:00
17/03/2006			0	0:00
18/03/2006			0	0:00
19/03/2006			0	0:00
20/03/2006			0	0:00
21/03/2006			0	0:00
22/03/2006			0	0:00
23/03/2006			0	0:00
24/03/2006			0	0:00
25/03/2006			0	0:00
26/03/2006			0	0:00
27/03/2006			0	0:00
28/03/2006			0	0:00
29/03/2006			0	0:00
30/03/2006			0	0:00
31/03/2006	4:00	5:00	1	1:00
31/03/2006	13:00	14:00	1	1:00
01/04/2006	4:00	5:00	1	1:00
02/04/2006			0	0:00
03/04/2006	11:00	14:00	1	3:00
03/04/2006	19:00	20:00	1	1:00
04/04/2006	21:00	22:00	1	1:00
05/04/2006	7:00	8:00	1	1:00
05/04/2006	15:00	16:00	1	1:00
06/04/2006			0	0:00
07/04/2006	6:00	7:00	1	1:00
08/04/2006			0	0:00
09/04/2006			0	0:00
10/04/2006			0	0:00
11/04/2006	7:00	8:00	1	1:00
11/04/2006	16:00	17:00	1	1:00
12/04/2006			0	0:00
13/04/2006	12:00	14:00	1	2:00
14/04/2006	5:00	6:00	1	1:00
14/04/2006	7:00	8:00	1	1:00
15/04/2006	17:00	19:00	1	2:00
16/04/2006	5:00	6:00	1	1:00
17/04/2006			0	0:00
18/04/2006	8:00	9:00	1	1:00
19/04/2006	13:00	14:00	1	1:00
20/04/2006			0	0:00
21/04/2006			0	0:00
22/04/2006	6:00	7:00	1	1:00
22/04/2006	9:00	10:00	1	1:00
22/04/2006	13:00	14:00	1	1:00

121

23/04/2006			0	0:00
24/04/2006			0	0:00
25/04/2006			0	0:00
26/04/2006	17:00	18:00	1	1:00
27/04/2006			0	0:00
28/04/2006			0	0:00
29/04/2006			0	0:00
30/04/2006			0	0:00
01/05/2006	3:00	4:00	1	1:00
01/05/2006	5:00	6:00	1	1:00
02/05/2006			0	0:00
03/05/2006	10:00	11:00	1	1:00
03/05/2006	15:00	16:00	1	1:00
04/05/2006	4:00	5:00	1	1:00
04/05/2006	20:00	21:00	1	1:00
05/05/2006	21:00	22:00	1	1:00
06/05/2006			0	0:00
07/05/2006			0	0:00
08/05/2006			0	0:00
09/05/2006			0	0:00
10/05/2006			0	0:00
11/05/2006			0	0:00
12/05/2006	18:00	19:00	1	1:00
13/05/2006			0	0:00
14/05/2006			0	0:00
15/05/2006	19:00	20:00	1	1:00
16/05/2006	10:00	16:00	1	6:00
17/05/2006	6:00	7:00	1	1:00
18/05/2006	12:00	13:00	1	1:00
19/05/2006			0	0:00
20/05/2006	18:00	19:00	1	1:00
21/05/2006			0	0:00
22/05/2006	16:00	17:00	1	1:00
23/05/2006			0	0:00
24/05/2006	6:00	7:00	1	1:00
25/05/2006			0	0:00
26/05/2006	8:00	9:00	1	1:00
27/05/2006			0	0:00
28/05/2006			0	0:00
29/05/2006	21:00	22:00	1	1:00
30/05/2006			0	0:00
31/05/2006	5:00	6:00	1	1:00
01/06/2006	17:00	18:00	1	1:00
02/06/2006	13:00	14:00	1	1:00
03/06/2006	0:00	19:00	1	19:00
04/06/2006			0	0:00

			0	0:00
05/06/2006			0	0:00
06/06/2006	2:00	6:00	1	4:00
07/06/2006			0	0:00
08/06/2006			0	0:00
09/06/2006			0	0:00
10/06/2006	18:00	19:00	1	1:00
11/06/2006			0	0:00
12/06/2006			0	0:00
13/06/2006			0	0:00
14/06/2006			0	0:00
15/06/2006			0	0:00
16/06/2006	18:00	19:00	1	1:00
17/06/2006	7:00	8:00	1	1:00
17/06/2006	14:00	15:00	1	1:00
18/06/2006			0	0:00
19/06/2006			0	0:00
20/06/2006			0	0:00
21/06/2006			0	0:00
22/06/2006	7:00	8:00	1	1:00
23/06/2006			0	0:00
24/06/2006			0	0:00
25/06/2006	5:00	6:00	1	1:00
26/06/2006	10:00	11:00	1	1:00
27/06/2006	1:00	2:00	1	1:00
28/06/2006	4:00	23:59	1	19:59
29/06/2006			0	0:00
30/06/2006	15:00	16:00	1	1:00
01/07/2006	8:00	9:00	1	1:00
02/07/2006			0	0:00
03/07/2006			0	0:00
04/07/2006			0	0:00
05/07/2006			0	0:00
06/07/2006			0	0:00
07/07/2006	23:00	23:59	1	0:59
08/07/2006	0:00	1:00	1	1:00
09/07/2006			0	0:00
10/07/2006			0	0:00
11/07/2006	9:00	13:00	1	4:00
12/07/2006			0	0:00
13/07/2006			0	0:00
14/07/2006			0	0:00
15/07/2006			0	0:00
16/07/2006			0	0:00
17/07/2006			0	0:00
18/07/2006			0	0:00
19/07/2006			0	0:00

20/07/2006	12:00	13:00	1	1:00
21/07/2006			0	0:00
22/07/2006			0	0:00
23/07/2006	13:00	16:00	1	3:00
24/07/2006			0	0:00
25/07/2006			0	0:00
26/07/2006			0	0:00
27/07/2006			0	0:00
28/07/2006			0	0:00
29/07/2006			0	0:00
30/07/2006			0	0:00
31/07/2006			0	0:00
01/08/2006			0	0:00
02/08/2006			0	0:00
03/08/2006	13:00	14:00	1	1:00
04/08/2006			0	0:00
05/08/2006			0	0:00
06/08/2006			0	0:00
07/08/2006	9:00	10:00	1	1:00
08/08/2006			0	0:00
09/08/2006	11:00	12:00	1	1:00
10/08/2006	19:00	20:00	1	1:00
11/08/2006			0	0:00
12/08/2006			0	0:00
13/08/2006	5:00	6:00	1	1:00
14/08/2006			0	0:00
15/08/2006			0	0:00
16/08/2006			0	0:00
17/08/2006	14:00	15:00	1	1:00
18/08/2006	9:00	10:00	1	1:00
19/08/2006			0	0:00
20/08/2006			0	0:00
21/08/2006			0	0:00
22/08/2006			0	0:00
23/08/2006			0	0:00
24/08/2006			0	0:00
25/08/2006	6:00	7:00	1	1:00
26/08/2006	10:00	11:00	1	1:00
27/08/2006			0	0:00
28/08/2006			0	0:00
29/08/2006			0	0:00
30/08/2006			0	0:00
31/08/2006			0	0:00
01/09/2006			0	0:00
02/09/2006			0	0:00
03/09/2006	2:00	3:00	1	1:00

Date	Start	End	Count	Duration
04/09/2006			0	0:00
05/09/2006			0	0:00
06/09/2006	2:00	3:00	1	1:00
06/09/2006	4:00	5:00	1	1:00
07/09/2006			0	0:00
08/09/2006			0	0:00
09/09/2006			0	0:00
10/09/2006			0	0:00
11/09/2006			0	0:00
12/09/2006			0	0:00
13/09/2006			0	0:00
14/09/2006			0	0:00
15/09/2006			0	0:00
16/09/2006			0	0:00
17/09/2006			0	0:00
18/09/2006	7:00	8:00	1	1:00
19/09/2006	16:00	17:00	1	1:00
20/09/2006			0	0:00
21/09/2006	6:00	7:00	1	1:00
21/09/2006	12:00	17:00	1	5:00
22/09/2006			0	0:00
23/09/2006	16:00	17:00	1	1:00
24/09/2006			0	0:00
25/09/2006			0	0:00
26/09/2006			0	0:00
27/09/2006			0	0:00
28/09/2006			0	0:00
29/09/2006			0	0:00
30/09/2006			0	0:00
01/10/2006			0	0:00
02/10/2006			0	0:00
03/10/2006			0	0:00
04/10/2006			0	0:00
05/10/2006			0	0:00
06/10/2006			0	0:00
07/10/2006			0	0:00
08/10/2006			0	0:00
09/10/2006			0	0:00
10/10/2006			0	0:00
11/10/2006			0	0:00
12/10/2006			0	0:00
13/10/2006			0	0:00
14/10/2006			0	0:00
15/10/2006			0	0:00
16/10/2006			0	0:00
17/10/2006			0	0:00

18/10/2006	13:00	15:00	1	2:00
19/10/2006			0	0:00
20/10/2006			0	0:00
21/10/2006			0	0:00
22/10/2006			0	0:00
23/10/2006			0	0:00
24/10/2006			0	0:00
25/10/2006			0	0:00
26/10/2006			0	0:00
27/10/2006			0	0:00
28/10/2006			0	0:00
29/10/2006			0	0:00
30/10/2006			0	0:00
31/10/2006	14:00	19:00	1	5:00
01/11/2006			0	0:00
02/11/2006			0	0:00
03/11/2006	9:00	14:00	1	5:00
03/11/2006	17:00	18:00	1	1:00
04/11/2006			0	0:00
05/11/2006			0	0:00
06/11/2006	18:00	19:00	1	1:00
07/11/2006	19:00	20:00	1	1:00
08/11/2006	5:00	6:00	1	1:00
09/11/2006	20:00	21:00	1	1:00
10/11/2006			0	0:00
11/11/2006			0	0:00
12/11/2006	17:00	18:00	1	1:00
13/11/2006			0	0:00
14/11/2006			0	0:00
15/11/2006			0	0:00
16/11/2006			0	0:00
17/11/2006			0	0:00
18/11/2006	12:00	13:00	1	1:00
18/11/2006	18:00	19:00	1	1:00
19/11/2006	19:00	20:00	1	1:00
20/11/2006			0	0:00
21/11/2006			0	0:00
22/11/2006			0	0:00
23/11/2006			0	0:00
24/11/2006			0	0:00
25/11/2006			0	0:00
26/11/2006			0	0:00
27/11/2006			0	0:00
28/11/2006			0	0:00
29/11/2006	12:00	13:00	1	1:00
29/11/2006	17:00	18:00	1	1:00

Date	Time Out	Time Restored	Frequency of outage	r (1 hour duration)
30/11/2006			0	0:00
01/12/2006			0	0:00
02/12/2006	14:00	16:00	1	2:00
03/12/2006	2:00	3:00	1	1:00
04/12/2006	22:00	23:00	1	1:00
05/12/2006	4:00	6:00	1	2:00
06/12/2006			0	0:00
07/12/2006			0	0:00
08/12/2006			0	0:00
09/12/2006			0	0:00
10/12/2006			0	0:00
11/12/2006			0	0:00
12/12/2006			0	0:00
13/12/2006			0	0:00
14/12/2006			0	0:00
15/12/2006			0	0:00
16/12/2006			0	0:00
17/12/2006			0	0:00
18/12/2006			0	0:00
19/12/2006			0	0:00
20/12/2006			0	0:00
21/12/2006			0	0:00
22/12/2006			0	0:00
23/12/2006			0	0:00
24/12/2006			0	0:00
25/12/2006			0	0:00
26/12/2006	16:00	18:00	1	2:00
27/12/2006			0	0:00
28/12/2006			0	0:00
29/12/2006			0	0:00
30/12/2006			0	0:00
31/12/2006			0	0:00
			108	180.97

Source: Archive of Power Holding Company of Nigeria (PHCN) 132 kV Switching Substation, Akure, Ondo State, Nigeria.

APPENDIX B.5: Data of Electric Power Outages in 2005

Summary of Outages in 2005				
Date	Time Out	Time Restored	Frequency of outage	r (1 hour duration)

01/01/2005			0	0:00
02/01/2005			0	0:00
03/01/2005			0	0:00
04/01/2005	9:00	10:00	1	1:00
05/01/2005			0	0:00
06/01/2005			0	0:00
07/01/2005			0	0:00
08/01/2005			0	0:00
09/01/2005			0	0:00
10/01/2005			0	0:00
11/01/2005			0	0:00
12/01/2005			0	0:00
13/01/2005			0	0:00
14/01/2005			0	0:00
15/01/2005			0	0:00
16/01/2005			0	0:00
17/01/2005			0	0:00
18/01/2005			0	0:00
19/01/2005			0	0:00
20/01/2005			0	0:00
21/01/2005			0	0:00
22/01/2005			0	0:00
23/01/2005			0	0:00
24/01/2005			0	0:00
25/01/2005			0	0:00
26/01/2005			0	0:00
27/01/2005			0	0:00
28/01/2005			0	0:00
29/01/2005			0	0:00
30/01/2005			0	0:00
31/01/2005			0	0:00
01/02/2005			0	0:00
02/02/2005			0	0:00
03/02/2005			0	0:00
04/02/2005			0	0:00
05/02/2005			0	0:00
06/02/2005			0	0:00
07/02/2005			0	0:00
08/02/2005			0	0:00
09/02/2005			0	0:00
10/02/2005			0	0:00
11/02/2005			0	0:00
12/02/2005			0	0:00
13/02/2005			0	0:00
14/02/2005	5:00	6:00	1	1:00

15/02/2005			0	0:00
16/02/2005			0	0:00
17/02/2005	10:00	11:00	1	1:00
18/02/2005			0	0:00
19/02/2005			0	0:00
20/02/2005			0	0:00
21/02/2005			0	0:00
22/02/2005			0	0:00
23/02/2005			0	0:00
24/02/2005	8:00	9:00	1	1:00
24/02/2005	10:00	11:00	1	1:00
24/02/2005	15:00	16:00	1	1:00
25/02/2005	14:00	15:00	1	1:00
25/02/2005	17:00	19:00	1	2:00
26/02/2005	12:00	13:00	1	1:00
27/02/2005	13:00	15:00	1	2:00
27/02/2005	19:00	20:00	1	1:00
28/02/2005	12:00	13:00	1	1:00
01/03/2005			0	0:00
02/03/2005			0	0:00
03/03/2005			0	0:00
04/03/2005			0	0:00
05/03/2005			0	0:00
06/03/2005			0	0:00
07/03/2005			0	0:00
08/03/2005	14:00	15:00	1	1:00
09/03/2005	16:00	17:00	1	1:00
10/03/2005			0	0:00
11/03/2005			0	0:00
12/03/2005	8:00	9:00	1	1:00
13/03/2005	19:00	20:00	1	1:00
14/03/2005			0	0:00
15/03/2005			0	0:00
16/03/2005			0	0:00
17/03/2005			0	0:00
18/03/2005			0	0:00
19/03/2005			0	0:00
20/03/2005			0	0:00
21/03/2005			0	0:00
22/03/2005	6:00	7:00	1	1:00
22/03/2005	11:00	12:00	1	1:00
22/03/2005	14:00	15:00	1	1:00
23/03/2005			0	0:00
24/03/2005	11:00	12:00	1	1:00
25/03/2005			0	0:00
26/03/2005	9:00	12:00	1	3:00

27/03/2005			0	0:00
28/03/2005			0	0:00
29/03/2005	12:00	16:00	1	4:00
30/03/2005			0	0:00
31/03/2005			0	0:00
01/04/2005			0	0:00
02/04/2005			0	0:00
03/04/2005			0	0:00
04/04/2005	14:00	15:00	1	1:00
04/04/2005	19:00	20:00	1	1:00
05/04/2005			0	0:00
06/04/2005	9:00	11:00	1	2:00
07/04/2005			0	0:00
08/04/2005			0	0:00
09/04/2005	3:00	4:00	1	1:00
10/04/2005			0	0:00
11/04/2005			0	0:00
12/04/2005			0	0:00
13/04/2005	9:00	10:00	1	1:00
14/04/2005			0	0:00
15/04/2005			0	0:00
16/04/2005			0	0:00
17/04/2005			0	0:00
18/04/2005			0	0:00
19/04/2005			0	0:00
20/04/2005			0	0:00
21/04/2005			0	0:00
22/04/2005			0	0:00
23/04/2005			0	0:00
24/04/2005			0	0:00
25/04/2005			0	0:00
26/04/2005	5:00	6:00	1	1:00
26/04/2005	17:00	18:00	1	1:00
27/04/2005			0	0:00
28/04/2005			0	0:00
29/04/2005			0	0:00
30/04/2005			0	0:00
01/05/2005			0	0:00
02/05/2005	18:00	19:00	1	1:00
03/05/2005			0	0:00
04/05/2005	18:00	19:00	1	1:00
05/05/2005	18:00	19:00	1	1:00
06/05/2005	19:00	20:00	1	1:00
07/05/2005			0	0:00
08/05/2005			0	0:00
09/05/2005			0	0:00

10/05/2005			0	0:00
11/05/2005	9:00	10:00	1	1:00
11/05/2005	21:00	22:00	1	1:00
12/05/2005	13:00	14:00	1	1:00
13/05/2005			0	0:00
14/05/2005	8:00	9:00	1	1:00
15/05/2005			0	0:00
16/05/2005			0	0:00
17/05/2005			0	0:00
18/05/2005			0	0:00
19/05/2005	14:00	15:00	1	1:00
20/05/2005	4:00	6:00	1	2:00
20/05/2005	20:00	21:00	1	1:00
21/05/2005	20:00	21:00	1	1:00
22/05/2005	13:00	14:00	1	1:00
22/05/2005	19:00	20:00	1	1:00
23/05/2005			0	0:00
24/05/2005			0	0:00
25/05/2005			0	0:00
26/05/2005	18:00	19:00	1	1:00
27/05/2005	5:00	6:00	1	1:00
27/05/2005	19:00	20:00	1	1:00
28/05/2005	6:00	7:00	1	1:00
29/05/2005			0	0:00
30/05/2005			0	0:00
31/05/2005			0	0:00
01/06/2005			0	0:00
02/06/2005			0	0:00
03/06/2005			0	0:00
04/06/2005			0	0:00
05/06/2005	1:00	2:00	1	1:00
06/06/2005	8:00	9:00	1	1:00
07/06/2005			0	0:00
08/06/2005			0	0:00
09/06/2005			0	0:00
10/06/2005	19:00	20:00	1	1:00
11/06/2005			0	0:00
12/06/2005			0	0:00
13/06/2005			0	0:00
14/06/2005	16:00	18:00	1	2:00
15/06/2005			0	0:00
16/06/2005			0	0:00
17/06/2005			0	0:00
18/06/2005			0	0:00
19/06/2005			0	0:00
20/06/2005			0	0:00

21/06/2005	19:00	20:00	1	1:00
22/06/2005	19:00	20:00	1	1:00
23/06/2005	8:00	9:00	1	1:00
23/06/2005	16:00	17:00	1	1:00
24/06/2005			0	0:00
25/06/2005	16:00	17:00	1	1:00
26/06/2005	14:00	19:00	1	5:00
27/06/2005			0	0:00
28/06/2005			0	0:00
29/06/2005			0	0:00
30/06/2005			0	0:00
01/07/2005			0	0:00
02/07/2005			0	0:00
03/07/2005			0	0:00
04/07/2005			0	0:00
05/07/2005			0	0:00
06/07/2005			0	0:00
07/07/2005			0	0:00
08/07/2005			0	0:00
09/07/2005			0	0:00
10/07/2005			0	0:00
11/07/2005			0	0:00
12/07/2005			0	0:00
13/07/2005			0	0:00
14/07/2005			0	0:00
15/07/2005			0	0:00
16/07/2005	10:00	16:00	1	6:00
17/07/2005			0	0:00
18/07/2005			0	0:00
19/07/2005			0	0:00
20/07/2005			0	0:00
21/07/2005			0	0:00
22/07/2005			0	0:00
23/07/2005			0	0:00
24/07/2005			0	0:00
25/07/2005			0	0:00
26/07/2005			0	0:00
27/07/2005			0	0:00
28/07/2005	10:00	11:00	1	1:00
28/07/2005	16:00	17:00	1	1:00
29/07/2005			0	0:00
30/07/2005			0	0:00
31/07/2005			0	0:00
01/08/2005			0	0:00
02/08/2005	11:00	12:00	1	1:00
02/08/2005	13:00	16:00	1	3:00

03/08/2005			0	0:00
04/08/2005			0	0:00
05/08/2005			0	0:00
06/08/2005			0	0:00
07/08/2005			0	0:00
08/08/2005			0	0:00
09/08/2005			0	0:00
10/08/2005			0	0:00
11/08/2005	22:00	23:00	1	1:00
12/08/2005			0	0:00
13/08/2005	14:00	15:00	1	1:00
14/08/2005			0	0:00
15/08/2005			0	0:00
16/08/2005			0	0:00
17/08/2005	23:00	23:59	1	0:59
18/08/2005	0:00	1:00	1	1:00
19/08/2005			0	0:00
20/08/2005	11:00	16:00	1	5:00
21/08/2005			0	0:00
22/08/2005			0	0:00
23/08/2005			0	0:00
24/08/2005	11:00	17:00	1	6:00
25/08/2005			0	0:00
26/08/2005	5:00	6:00	1	1:00
27/08/2005			0	0:00
28/08/2005			0	0:00
29/08/2005			0	0:00
30/08/2005			0	0:00
31/08/2005			0	0:00
01/09/2005			0	0:00
02/09/2005			0	0:00
03/09/2005			0	0:00
04/09/2005			0	0:00
05/09/2005			0	0:00
06/09/2005	8:00	9:00	1	1:00
07/09/2005			0	0:00
08/09/2005			0	0:00
09/09/2005			0	0:00
10/09/2005			0	0:00
11/09/2005			0	0:00
12/09/2005	1:00	5:00	1	4:00
13/09/2005			0	0:00
14/09/2005			0	0:00
15/09/2005			0	0:00
16/09/2005			0	0:00
17/09/2005	10:00	11:00	1	1:00

17/09/2005	12:00	13:00	1	1:00
18/09/2005	10:00	15:00	1	5:00
19/09/2005			0	0:00
20/09/2005	13:00	14:00	1	1:00
21/09/2005			0	0:00
22/09/2005			0	0:00
23/09/2005			0	0:00
24/09/2005	19:00	20:00	1	1:00
25/09/2005			0	0:00
26/09/2005			0	0:00
27/09/2005			0	0:00
28/09/2005			0	0:00
29/09/2005			0	0:00
30/09/2005			0	0:00
01/10/2005			0	0:00
02/10/2005			0	0:00
03/10/2005			0	0:00
04/10/2005	11:00	12:00	1	1:00
04/10/2005	15:00	16:00	1	1:00
05/10/2005			0	0:00
06/10/2005			0	0:00
07/10/2005			0	0:00
08/10/2005			0	0:00
09/10/2005			0	0:00
10/10/2005			0	0:00
11/10/2005	14:00	15:00	1	1:00
12/10/2005			0	0:00
13/10/2005			0	0:00
14/10/2005			0	0:00
15/10/2005	9:00	10:00	1	1:00
15/10/2005	19:00	20:00	1	1:00
15/10/2005	20:00	21:00	1	1:00
16/10/2005			0	0:00
17/10/2005			0	0:00
18/10/2005			0	0:00
19/10/2005			0	0:00
20/10/2005			0	0:00
21/10/2005			0	0:00
22/10/2005			0	0:00
23/10/2005			0	0:00
24/10/2005			0	0:00
25/10/2005			0	0:00
26/10/2005	9:00	10:00	1	1:00
27/10/2005			0	0:00
28/10/2005			0	0:00
29/10/2005			0	0:00

30/10/2005	19:00	20:00	1	1:00
31/10/2005			0	0:00
01/11/2005			0	0:00
02/11/2005			0	0:00
03/11/2005			0	0:00
04/11/2005			0	0:00
05/11/2005			0	0:00
06/11/2005			0	0:00
07/11/2005			0	0:00
08/11/2005			0	0:00
09/11/2005			0	0:00
10/11/2005			0	0:00
11/11/2005			0	0:00
12/11/2005			0	0:00
13/11/2005			0	0:00
14/11/2005			0	0:00
15/11/2005			0	0:00
16/11/2005			0	0:00
17/11/2005			0	0:00
18/11/2005			0	0:00
19/11/2005			0	0:00
20/11/2005			0	0:00
21/11/2005			0	0:00
22/11/2005			0	0:00
23/11/2005			0	0:00
24/11/2005	15:00	16:00	1	1:00
25/11/2005	9:00	10:00	1	1:00
26/11/2005	6:00	7:00	1	1:00
27/11/2005			0	0:00
28/11/2005			0	0:00
29/11/2005			0	0:00
30/11/2005	2:00	4:00	1	2:00
01/12/2005			0	0:00
02/12/2005			0	0:00
03/12/2005			0	0:00
04/12/2005			0	0:00
05/12/2005			0	0:00
06/12/2005			0	0:00
07/12/2005			0	0:00
08/12/2005			0	0:00
09/12/2005			0	0:00
10/12/2005			0	0:00
11/12/2005			0	0:00
12/12/2005			0	0:00
13/12/2005			0	0:00
14/12/2005			0	0:00

15/12/2005			0	0:00
16/12/2005			0	0:00
17/12/2005			0	0:00
18/12/2005			0	0:00
19/12/2005			0	0:00
20/12/2005			0	0:00
21/12/2005			0	0:00
22/12/2005			0	0:00
23/12/2005			0	0:00
24/12/2005			0	0:00
25/12/2005			0	0:00
26/12/2005			0	0:00
27/12/2005			0	0:00
28/12/2005			0	0:00
29/12/2005			0	0:00
30/12/2005			0	0:00
31/12/2005	16:00	17:00	1	1:00
			89	126.98

Source: Archive of Power Holding Company of Nigeria (PHCN) 132 kV Switching
Substation, Akure, Ondo State, Nigeria.

APPENDIX C

TABLE C.1: Power Holding Company of Nigeria Average Customer Population for 2009

Power Holding Company of Nigeria Average Customer Population for 2009						
Customer Population by Tariff Classification		Customer Type	Inactive	Active	Total	Energy Delivered (kWh)
Residential	37,476	Maximum Demand	62	406	468	
Commercial	6,892	Prime	113	570	683	
Industrial	372	Non-Maximum Demand	3,073	43,789	46862	
Street Lighting	66	Prepaid Meter	-	41	-	
Sub-total	44,806	Total	3,248	44,806	48,054	15,845,400

TABLE C.2: Power Holding Company of Nigeria Average Customer Population for 2008

Power Holding Company of Nigeria Average Customer Population for 2008						
Customer Population by Tariff Classification		Customer Type	Inactive	Active	Total	Energy Delivered (kWh)
Residential	37,432	Maximum Demand	56	389	445	
Commercial	6,860	Prime	107	537	644	
Industrial	353	Non-Maximum Demand	3,058	43,752	46810	
Street Lighting	53	Prepaid Meter	-	20	-	
Sub-total	44,698	Total	3,221	44,698	47,919	15,844,277

137

TABLE C.3: Power Holding Company of Nigeria Average Customer Population for 2007

Power Holding Company of Nigeria Average Customer Population for 2007						
Customer Population by Tariff Classification		Customer Type	Inactive	Active	Total	Energy Delivered (kWh)
Residential	37,405	Maximum Demand	51	368	419	
Commercial	6,839	Prime	101	511	612	
Industrial	341	Non-Maximum Demand	3,025	43,738	46763	
Street Lighting	44	Prepaid Meter	-	12	-	
Sub-total	44,629	Total	3,177	44,629	47,806	15,843,176

TABLE C.4: Power Holding Company of Nigeria Average Customer Population for 2006

Power Holding Company of Nigeria Average Customer Population for 2006						
Customer Population by Tariff Classification		Customer Type	Inactive	Active	Total	Energy Delivered (kWh)
Residential	37,381	Maximum Demand	48	349	397	
Commercial	6,823	Prime	98	499	597	
Industrial	334	Non-Maximum Demand	3,007	43,724	46731	
Street Lighting	38	Prepaid Meter	-	4	-	
Sub-total	44,576	Total	3,153	44,576	47,729	15,842,194

TABLE C.5: Power Holding Company of Nigeria Average Customer Population for 2005

Power Holding Company of Nigeria Average Customer Population for 2005						
Customer Population by Tariff Classification		Customer Type	Inactive	Active	Total	Energy Delivered (kWh)
Residential	37,362	Maximum Demand	46	338	384	
Commercial	6,810	Prime	95	487	582	
Industrial	331	Non-Maximum Demand	2,998	43,713	46711	
Street Lighting	35	Prepaid Meter	-	-	-	
Sub-total	44,538	Total	3,139	44,538	47,677	15,841,234

Source: Power Holding Company of Nigeria, Akure Business Unit.

139

Printed by Books on Demand GmbH, Norderstedt / Germany